BIOFLUID
MECHANICS

T0320618

BIOFLUID MECHANICS

Jagan N. Mazumdar

Department of Applied Mathematics
The University of Adelaide
South Australia

World Scientific
Singapore • New Jersey • London • Hong Kong

Published by

World Scientific Publishing Co. Pte. Ltd.
P O Box 128, Farrer Road, Singapore 9128
USA office: Suite 1B, 1060 Main Street, River Edge, NJ 07661
UK office: 73 Lynton Mead, Totteridge, London N20 8DH

BIOFLUID MECHANICS

ISBN 981-02-0927-4

Printed in Singapore by JBW Printers & Binders Pte. Ltd.

Dedicated to
My Beloved Mother
and to
My Beloved Guru
Bhagawan Sri Sathya Sai Baba

PREFACE

For more than a decade I have been involved in studying various aspects of biofluid mechanics. During that period, I have been fortunate to have had a number of excellent collaborators as graduate students and research workers from whom I learned a lot on this subject. In preparing this monograph I have borrowed substantially from the work of these collaborators and I have frequently quoted from their publications.

The present book can be regarded as a complement to my previous book *An Introduction to Mathematical Physiology and Biology* published by the Cambridge University Press in 1989. Readers who are familiar with my previous book, will find the present book a natural continuation of the previous one.

The book gives a systematic account of a number of topics commonly discussed under physiological and biological fluid dynamics. It begins with an introduction to elementary fluid mechanics as taught at undergraduate level. The basic equations of motion for both viscous and inviscid fluids are given in detail. This is followed by the mechanics of flow in the circulatory system. Circulatory mechanics is a wide subject and in the present book only systematic and pulmonary circulations are explained. In Chapter 3 flow properties of blood are presented. A brief mention of some diseases related to obstruction of blood flow is made. Chapter 4 discusses some well-known models of biofluid flows. Applications to duct and pipe flow are also given. Non-Newtonian fluid flow is discussed in Chapter 5. Some models of other flows are included in Chapter 6, ie., flows in capillary blood vessels, in kidneys and the peristaltic flows. The fluid mechanics of heart valves, natural and prosthetic, are discussed in Chapter 7. The last chapter deals with the computational methods in biofluid mechanics.

In preparing an account of biofluid mechanics, however, there may be difficulties in finding the right balance between biological and physiological fluid motion on one side and mathematical modelling on the other side. The balance has been given in the present book.

Some of the topics presented in the monograph are old and well-known, but some are relatively new. Although I wanted to include a much wider selection of topics, I found that this was not possible because of the degree of thoroughness that would have been required. I therefore wish to apologize in advance for this limitation.

There are many people whose help I must acknowledge. I wish to mention especially my former students Kym Thalassoudis, Lai-Leng Soh and Keng Cheng Ang whose research have given me motivation in writing this book. I am also thankful to Miss Danielle Hopkins for meticulously typesetting the manuscript using LaTeX and to Mrs Jenny Laing for her excellent care and cooperation in drawing the figures for the book. Finally, I wish to express my gratitude to my wife, Maya, for the unlimited personal encouragement, patience and support I received during the course of writing this monograph and without which it would never have been finished.

In spite of great care and precautions, I am sure that many mistakes and defects remain in the book. I would be most grateful if these are brought to my attention so that they could be attended in the future.

J. MAZUMDAR
ADELAIDE, AUSTRALIA

CONTENTS

BIOFLUID
MECHANICS

Chapter 1

INTRODUCTION

1.1 A Basic Introduction

Biological Science has long been advanced from purely descriptive to analytical science. Many analytical methods of physical science have been used successively in the study of biological science. Biofluid mechanics is the study of certain class of biological problems from fluid mechanics points of view. Biofluid mechanics does not involves any new development of the general principles of fluid mechanics but it involves some new applications of the method of fluid mechanics. Complex movements of fluids in the biological system demand for their analysis professional fluid mechanics skills. The most common fluid mechanics problem in the biological system is the flow of blood. In 1840, the French physician Poiseuille was interested in the flow of blood and conducted a study of flow in capillaries. It is the well-known Poiseuille flow in fluid mechanics, but we know that ordinary Poiseuille flow does not represent the actual blood flow in a biological system. Many biofluid mechanics problems are concerned with classical fluid mechanics, but also its modern aspects such as rheology, chemical reactions, electrothermal effects etc. Approaching the problem from a macroscopic (continuum approach) point of view by assuming that any small volume element of the fluid is so large that it contains a very large number of molecules and also assuming that the fluid is a continuous medium, the dependent variables describing the fluid motion are as follows:

$$\text{pressure } p = p(x, y, z, t),$$
$$\text{density } \rho = \rho(x, y, z, t),$$
$$\text{velocity } \mathbf{q} = \mathbf{q}(x, y, z, t) = (u, v, w).$$

1

The number of basic variables in fluid mechanics is five: three velocity components and two thermodynamic properties. The flow field is completely determined once we specify the velocity vector **q** and two thermodynamic properties as a function of space and time. Hence we need five independent equations. These are usually the three components of the equation of motion, a continuity equation and an energy equation. We will discuss this in the next section.

If the density ρ is constant (does not vary with pressure) throughout then the flow is said to be *incompressible*. In fluid mechanics, it is customary to treat liquids as incompressible fluids. In an incompressible fluid the energy equation is not needed since density is taken as known and only pressure together with velocity need to be found in order to completely describe the flow of fluids.

In general, we divide fluids into two classes — viscous and non-viscous (inviscid). A brief description of viscosity will now be given for the benefit of the reader.

1.1.1 *Viscosity*

Newton defined the viscosity of a fluid as a lack of slipperiness between the layers of the fluid and, of course, in doing so he implied that there was such a thing as a "layer of fluid" or "laminae" of the fluid, and the viscosity arises as a result of rubbing one lamina upon the other.

Consider two laminae which are in contact with one another (see Fig. 1.1).

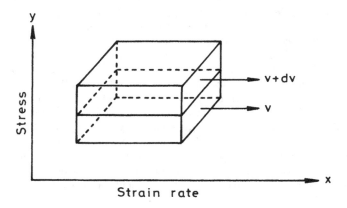

Fig. 1.1: Newtonian concept of viscosity.

Suppose some force F parallel to the x-axis acts and produces relative motion between the lamina, ie., the top lamina moves with velocity dv relative to the bottom lamina. Hence, there is a rate of change of velocity with distance in the y direction, ie., there exists a velocity gradient $\frac{dv}{dy}$. It is hypothesized that

$$F \propto A\frac{dv}{dy} \qquad (1.1)$$

where A is the area of contact between the laminae. The proportionality constant is then defined to be the viscosity of the fluid, and is usually denoted by μ, ie.,

$$F = \mu A\frac{dv}{dy}. \qquad (1.2)$$

The dimensions of the viscosity are given by

$$[\mu] = \frac{[\text{force}]}{[\text{area}]\frac{[\text{velocity}]}{[\text{length}]}} = \frac{MLT^{-2}}{L^2\frac{LT^{-1}}{L}} = \frac{M}{LT}. \qquad (1.3)$$

It is customary to classify fluids as *Ideal* or *non-Viscous* (μ=0), *Newtonian* (μ is constant), or *non-Newtonian* (μ is variable). Actually, no fluid is ever really ideal, but some fluids, at least in certain regions of flow under certain circumstances, approach ideal conditions very closely.

1.1.2 *Laminar and Turbulent Flow*

The terms laminar flow and purely viscous flow are used synonymously to mean a fluid flow which flows in laminae, as opposed to turbulent flow in which the velocity components have random fluctuations imposed upon their mean values. For a given fluid, as the velocity increases the flow changes from laminar to turbulent, passing through a transition region. Both laminar and turbulent flow occur in nature, but turbulent flow seems to be a more natural state of affairs.

1.1.3 *Compressible and Incompressible Flow*

It is common practice to divide fluids into two groups — liquids and gases. Gases are compressible and their density changes readily with temperature and pressure. Liquids, on the other hand, are rather difficult to compress and for all practical purposes may be considered as incompressible.

From above it can be seen that there are many different properties that a fluid may exhibit. However, the theoretical development of fluid mechanics can primarily be broken down into two classes:

(i) Inviscid fluid mechanics (ie., non-viscous or viscosity can be neglected)

(ii) Viscous fluid mechanics

In the next section, we give the basic equations of fluid mechanics for application in subsequent sections.

1.2 Basic Equations of Fluid Mechanics

As mentioned earlier, there are five basic variables in fluid flow, three components of velocity and two thermodynamic properties. Hence there are five basic equations which describe the flow; the continuity equation, three components of the momentum equation, and the energy equation. In general, the energy equation becomes uncoupled in *incompressible* flow since the density is constant. In turbulent flow the situation is somewhat more complex and a closed set of equations cannot generally be developed.

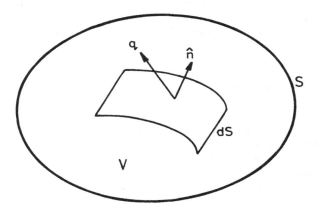

Fig. 1.2: Volume V enclosed by surface S.

1.2.1 *Continuity Equation*

Consider a fixed volume V in space in our flow field with closed surface S (see Fig. 1.2). Assuming there are no sources or sinks in V and applying the Principle of Conservation of Mass, that is,

Time rate of change of total mass in V
= Rate of inflow (or outflow) of mass across S

we get:

$$\frac{\partial}{\partial t} \int_V \rho \, dV \;=\; \int_S -\rho \, \mathbf{q}.\hat{\mathbf{n}} \, dS \tag{1.4}$$

where ρ = density at a point in V at time t
 $\mathbf{q}$ = velocity at a point in V at time t
 $\hat{\mathbf{n}}$ = outward unit normal to the surface at a point.

The fact that the density of a fluid can vary from point to point makes the fluid compressible.

Using Gauss' Divergence Theorem, that is

$$\int_S \mathbf{F}.\hat{\mathbf{n}} \, dS \;=\; \int_V div\mathbf{F} \, dV \tag{1.5}$$

Eq. 1.4 becomes

$$\int_V \frac{\partial \rho}{\partial t} \, dV \;=\; \int_V -div(\rho \, \mathbf{q}) \, dV$$
$$\int_V \left[\frac{\partial \rho}{\partial t} + div(\rho \, \mathbf{q}) \right] \, dV = 0. \tag{1.6}$$

As the choice of V is arbitrary,

$$\frac{\partial \rho}{\partial t} + div(\rho \, \mathbf{q}) \;=\; 0$$
$$\frac{\partial \rho}{\partial t} + \mathbf{q}.\nabla \rho + \rho \, div\mathbf{q} \;=\; 0 \tag{1.7}$$
$$\frac{D\rho}{Dt} + \rho \, div\mathbf{q} \;=\; 0$$

where the material derivative is

$$\frac{D}{Dt} \;\equiv\; \frac{\partial}{\partial t} + \mathbf{q}.\nabla. \tag{1.8}$$

Equation 1.7 is called the *Conservation of Mass equation* or the *Continuity equation*. It is widely applicable as no special assumption has been made so far as to what type of fluid can be considered.

In the case of an incompressible fluid, we have

$$\frac{D\rho}{Dt} \;=\; 0. \tag{1.9}$$

Hence, the continuity equation becomes

$$div\mathbf{q} = 0 \qquad (1.10)$$

ie.,

$$\frac{\partial u}{\partial x} + \frac{\partial v}{\partial y} + \frac{\partial w}{\partial z} = 0 \qquad (1.11)$$

where u, v and w denote velocity components in the x, y and z directions respectively.

1.2.2 Equations of Motion

In a fluid, forces can be divided into two types, namely, external forces (viz. gravitational, body forces etc.) and internal forces.

External forces are forces acting on a local portion of fluid due to some agency outside the fluid. It is expressed as a force per unit mass, $\mathbf{F}$. So in a volume δV of the fluid, we have mass $\rho\delta V$ with associated external or body force $\mathbf{F}\,\rho\delta V$.

Internal forces are forces due to interaction between neighbouring fluid particles. These are expressed as forces per unit area or *stresses* and are exerted across the boundary surface between elements.

Let S be a surface between the fluid considered and the 'external' fluid (can be a fluid of the same type) and dS be an element of S with unit normal $\hat{n}$. Also let $\mathbf{T}$ be the stress vector (force per unit area) giving the force between the internal and external fluids. Then the internal forces acting between sections of fluid for the surface element dS is

$$(\mathbf{T}.\hat{n})\,dS\,\hat{n} = \mathbf{T}^n dS \quad \text{(say)}. \qquad (1.12)$$

In a three-dimensional case, a full description of these forces requires vectors in three mutually perpendicular directions, that is, we need a nine-component entity — a tensor $\sigma_{ij}, (i, j = 1, 2, 3)$. (σ_{ij}) is called the *stress tensor* and σ_{ij} corresponds to the force component per unit area. The first subscript denotes the face on which the stress acts, and the second subscript denotes the direction of stress. The face is indicated as the plane perpendicular to the axis of the subscript. For instance, the 3rd face is perpendicular to the z or x_3 axis (see Fig. 1.3).

In tensor notation,

$$\begin{aligned}
\mathbf{T}^n dS &= \mathbf{T}^j n_j dS \\
\text{or,} \quad T_i^n dS &= T_i^j n_j dS \\
&= \sigma_{ij} n_j\, dS \\
&= \sigma_{ij}\, dS_j.
\end{aligned} \qquad (1.13)$$

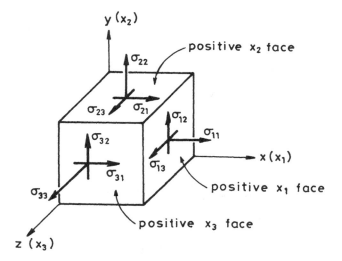

Fig. 1.3: Stresses at a point in space. The positive faces are shown; the opposite ones are negative faces.

Let us now consider a viscous fluid. The inviscid case can be obtained simply by putting the viscosity, μ, to be zero.

Assuming symmetry of the stress tensor, that is,

$$\sigma_{ij} = \sigma_{ji} \tag{1.14}$$

and using the Newtonian relation for the viscous part

$$\text{stress} = \text{viscosity} \times \text{rate of strain},$$

we obtain

$$\sigma_{ij} = \underbrace{-p\,\delta_{ij}}_{inviscid\,part} + \underbrace{\mu \left(\frac{\partial u_i}{\partial x_j} + \frac{\partial u_j}{\partial x_i} \right)}_{viscous\,part} \tag{1.15}$$

where u_i = velocity of the fluid flow in the x_i-direction
$\quad\;\; p$ = pressure acting on the surface (hydrostatic pressure)
$\quad\;\; \mu$ = coefficient of viscosity

$\delta_{ij} = \begin{cases} 1 & \text{if } i = j \\ 0 & \text{otherwise.} \end{cases}$

Consider a general fluid volume V with bounding surface S as in Fig. 1.2. Since,

Time rate of change of momentum = Net force acting on fluid

we have

$$\frac{D}{Dt}\int_V \rho\, \mathbf{q}\, dV = \underbrace{\int_V \rho\, \mathbf{F}\, dV}_{net\ body\ force} + \underbrace{\int_S \mathbf{T}^n dS}_{net\ surface\ force\ due\ to\ external\ fluid} \tag{1.16}$$

Using Reynolds Transport theorem, which states that

$$\frac{D}{Dt}\int_V G\, dV = \int_V \left(\frac{DG}{Dt} + G\, div\mathbf{q}\right)dV \tag{1.17}$$

we obtain from Eq. 1.16

$$\int_V \left[\frac{D}{Dt}(\rho\mathbf{q}) + \rho\mathbf{q}\, div\mathbf{q}\right]dV = \int_V \rho\mathbf{F}\, dV + \int_S \mathbf{T}^n dS.$$

For an incompressible fluid, the above becomes

$$\int_V \rho\frac{D\mathbf{q}}{Dt}\, dV = \int_V \rho\mathbf{F}\, dV + \int_S \mathbf{T}^n\, dS \tag{1.18}$$

or in indicial notation,

$$\int_V \rho\frac{Dq_i}{Dt}\, dV = \int_V \rho F_i\, dV + \int_S T_i^n\, dS. \tag{1.19}$$

Substituting Eq. 1.13 into Eq. 1.19,

$$\int_V \rho\frac{Dq_i}{Dt}\, dV = \int_V \rho F_i\, dV + \int_S \sigma_{ij}\, dS_j. \tag{1.20}$$

Applying Gauss' Divergence theorem, this becomes

$$\int_V \rho\frac{Dq_i}{Dt}\, dV = \int_V \rho F_i\, dV + \int_V \frac{\partial \sigma_{ij}}{\partial x_j}\, dV. \tag{1.21}$$

As the choice of V is arbitrary, we get

$$\rho\frac{Dq_i}{Dt} = \rho F_i + \frac{\partial \sigma_{ij}}{\partial x_j}. \tag{1.22}$$

From Eq. 1.15,

$$\begin{aligned}
\frac{\partial \sigma_{ij}}{\partial x_j} &= -\frac{\partial p}{\partial x_j}\delta_{ij} + \mu\frac{\partial^2 u_i}{\partial x_j \partial x_j} + \mu\frac{\partial^2 u_j}{\partial x_j \partial x_i} \\
&= -\frac{\partial p}{\partial x_i} + \mu\nabla^2 u_i + \mu\frac{\partial}{\partial x_i}\left(\frac{\partial u_j}{\partial x_j}\right) \\
&= -\frac{\partial p}{\partial x_i} + \mu\nabla^2 u_i + \mu\frac{\partial}{\partial x_i}(div\mathbf{q}).
\end{aligned} \tag{1.23}$$

In the case of an incompressible fluid, $div\mathbf{q} = 0$. Thus

$$\frac{\partial \sigma_{ij}}{\partial x_j} = -\nabla p|_i + \mu \nabla^2 u_i. \tag{1.24}$$

Substituting this into Eq. 1.22 we get

$$\rho \frac{D\mathbf{q}}{Dt} = \rho \mathbf{F} - \nabla p + \mu \nabla^2 \mathbf{q}. \tag{1.25}$$

This is known as the *Navier-Stokes equation* for a viscous, incompressible fluid. In the Cartesian coordinate system this reduces to

$$\rho \left(\frac{\partial u}{\partial t} + u \frac{\partial u}{\partial x} + v \frac{\partial u}{\partial y} + w \frac{\partial u}{\partial z} \right) = \rho F_x - \frac{\partial p}{\partial x} + \mu \nabla^2 u \tag{1.26}$$

$$\rho \left(\frac{\partial v}{\partial t} + u \frac{\partial v}{\partial x} + v \frac{\partial v}{\partial y} + w \frac{\partial v}{\partial z} \right) = \rho F_y - \frac{\partial p}{\partial y} + \mu \nabla^2 v \tag{1.27}$$

$$\rho \left(\frac{\partial w}{\partial t} + u \frac{\partial w}{\partial x} + v \frac{\partial w}{\partial y} + w \frac{\partial w}{\partial z} \right) = \rho F_z - \frac{\partial p}{\partial z} + \mu \nabla^2 w. \tag{1.28}$$

If however, the flow is frictionless then there are no shear stresses and the normal stresses are just the pressure which is isotropic ie.,

$$\sigma_{ij} = -p \, \delta_{ij}. \tag{1.29}$$

In this case the equations of motion become

$$\rho \frac{D\mathbf{q}}{Dt} = \rho \mathbf{F} - \nabla p. \tag{1.30}$$

These equations are known as the Euler equations for frictionless flow. If the flow is incompressible, the density is constant and the three Euler equations (1.30) plus the continuity equation (1.10) constitute four equations with four unknowns u, v, w, and p.

If we compare the complete Navier-Stokes equations (1.25) with the Euler equations (1.30), we notice that the differences are the additional terms involving the viscosity. However, if the fluid is incompressible, even if the viscosity is not zero, the viscous terms in Eq. 1.25 are negligible if derivatives of the form $\frac{\partial u_i}{\partial x_j}$ are small, and the Navier-Stokes equations reduce to the Euler equations. Thus, the Navier-Stokes equations reduce to the Euler equations if the viscosity is negligible or the derivatives of velocities are negligible. Indeed, the last conclusion is very important in fluid mechanics because for many fluid flows it is assumed that there are two regions: one close to a solid boundary where the viscous terms are important and thus the Navier-Stokes equations are considered, and a region away from the boundary where

Euler's equations hold good approximately. The flows of fluid in these two regions are referred to as "Boundary Layer Flow" and "Potential Flow". We will discuss these in more detail in subsequent chapters.

1.3 Simplification of Basic Equations

The continuity equation and the equations of motion can be simplified in the following cases:

(i)   When there are *no external body forces*, ie., when $F_x = F_y = F_z = 0$, or when the external forces form a conservative system. If the external forces form a conservative system, then they can be derived as the gradient of a scalar potential function Ω such that

$$\rho\,F_x = -\frac{\partial\Omega}{\partial x}, \quad \rho\,F_y = -\frac{\partial\Omega}{\partial y}, \quad \rho\,F_z = -\frac{\partial\Omega}{\partial z}. \tag{1.31}$$

In this case the terms $(\rho\,\mathbf{F} - \nabla p)$ appearing in Eq. 1.25 become

$$\rho\mathbf{F} - \nabla p \;=\; -\nabla(p + \Omega) \tag{1.32}$$

so that the pressure is effectively replaced by $p + \Omega$.

(ii) When the flow is *steady*, ie., when the flow does not vary with time so that u, v, w and p are functions of x, y, z only and $\frac{\partial u}{\partial t}, \frac{\partial v}{\partial t}, \frac{\partial w}{\partial t}$ and $\frac{\partial p}{\partial t}$ are all zero. If the flow is such that the velocity components and thermodynamic properties at every point in space change in time, the flow is called *unsteady*.

(iii) When the flow is *two-dimensional*, ie., when the flow is the same in all planes parallel to say, the xy-plane. In this case we assume that $w = 0$ and there is no variation with respect to z. In such a case we may think of a volume of fluid of unit thickness moving parallel to the xy-plane. Thus, the continuity equation (1.11) and the Navier-Stokes equations (1.26)-(1.27) in the absence of external body forces take the form

$$\frac{\partial u}{\partial x} + \frac{\partial v}{\partial y} \;=\; 0 \tag{1.33}$$

$$\rho\left(\frac{\partial u}{\partial t} + u\,\frac{\partial u}{\partial x} + v\,\frac{\partial u}{\partial y}\right) \;=\; -\frac{\partial p}{\partial x} + \mu\left(\frac{\partial^2 u}{\partial x^2} + \frac{\partial^2 u}{\partial y^2}\right) \tag{1.34}$$

$$\rho\left(\frac{\partial v}{\partial t} + u\,\frac{\partial v}{\partial x} + v\,\frac{\partial v}{\partial y}\right) \;=\; -\frac{\partial p}{\partial y} + \mu\left(\frac{\partial^2 v}{\partial x^2} + \frac{\partial^2 v}{\partial y^2}\right). \tag{1.35}$$

We note that Eq. 1.33 is identically satisfied if we assume a function $\psi(x,y)$ such that

$$u = \frac{\partial \psi}{\partial y}, \quad v = -\frac{\partial \psi}{\partial x}. \tag{1.36}$$

The function $\psi(x,y)$ is called the *Stream function* since $\psi(x,y) = const$ gives us a streamline for the flow. Clearly, from Eq. 1.36 we note that the directional derivative of ψ in any direction gives the component of the velocity in a direction making an angle $-90°$ with the direction of the derivative. Since the velocity vector $\mathbf{q}$ is tangent to a streamline, it has no component normal to the streamline, hence $\frac{\partial \psi}{\partial s} = 0$ where s is measured along the streamline, which implies that $\psi = const$ along each streamline. Obviously, the value of ψ will vary from streamline to streamline. If we define dn as the element in a direction normal to ds and if the angle between ds and dn is $90°$ then we have

$$\frac{\partial \psi}{\partial n} = q \quad \text{and} \quad d\psi = q \, dn. \tag{1.37}$$

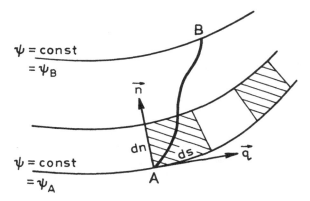

Fig. 1.4: The streamlines and the stream function ψ.

Thus, the increment $d\psi$ in ψ from one streamline to the next equals the amount of fluid passing between these two streamlines per unit time. It therefore follows that the amount of fluid crossing any curve AB (Fig. 1.4) in unit time is $\psi_B - \psi_A$, which is the value of ψ at B minus the value of ψ at A. Using the stream function ψ, Eqs. 1.34 and 1.35 can be combined and after eliminating p between them transform into

$$\frac{\partial}{\partial t}\nabla^2\psi + \frac{\partial\psi}{\partial y}\frac{\partial}{\partial x}\nabla^2\psi - \frac{\partial\psi}{\partial x}\frac{\partial}{\partial y}\nabla^2\psi = \nu\nabla^4\psi \tag{1.38}$$

where ν is the kinematic viscosity μ/ρ. If further we introduce the *vorticity* of this two-dimensional flow as

$$\omega = \frac{1}{2}\left(\frac{\partial v}{\partial x} - \frac{\partial u}{\partial y}\right) = -\frac{1}{2}\nabla^2\psi \tag{1.39}$$

then Eq. 1.38 reduces to

$$\frac{\partial\omega}{\partial t} + \frac{\partial\psi}{\partial y}\frac{\partial\omega}{\partial x} - \frac{\partial\psi}{\partial x}\frac{\partial\omega}{\partial y} = \nu\nabla^2\omega. \tag{1.40}$$

(iv) When the flow is *one-dimensional* ie., if there is only one non-zero velocity component, say u, then the continuity equation reduces to

$$\frac{\partial\rho}{\partial t} + \frac{\partial}{\partial x}(\rho u) = 0 \tag{1.41}$$

or, if the fluid is incompressible

$$\frac{\partial u}{\partial x} = 0 \tag{1.42}$$

and the Navier-Stokes equations become a single equation

$$\rho\frac{\partial u}{\partial t} = -\frac{\partial p}{\partial x} + \mu\left(\frac{\partial^2 u}{\partial y^2} + \frac{\partial^2 u}{\partial z^2}\right) + \rho F_x. \tag{1.43}$$

(v) When the flow is *axially symmetric*, that is, when the flow is symmetric about an axis, further simplification of the basic equations can be carried out. If we use cylindrical polar coordinates (r, θ, z), where the axis of symmetry is along the axis of z, then in general we have three components of velocity; v_r along the radius vector perpendicular to the axis, v_θ perpendicular to the axis and the radius vector, and v_z parallel to the axis of z. In the axi-symmetric case, we take $v_\theta = 0$, and v_r, v_z and p to be independent of θ. Hence, the continuity equation and equations of motion become

$$\frac{1}{r}\frac{\partial}{\partial r}(rv_r) + \frac{\partial}{\partial z}v_z = 0, \tag{1.44}$$

$$\rho\left(\frac{\partial v_r}{\partial t} + v_r\frac{\partial v_r}{\partial r} + v_z\frac{\partial v_r}{\partial z}\right) = -\frac{\partial p}{\partial r} + \mu\left(\frac{\partial^2 v_r}{\partial r^2} + \frac{1}{r}\frac{\partial v_r}{\partial r} - \frac{v_r}{r^2} + \frac{\partial^2 v_r}{\partial z^2}\right), \quad (1.45)$$

$$\rho\left(\frac{\partial v_z}{\partial t} + v_r\frac{\partial v_z}{\partial r} + v_z\frac{\partial v_z}{\partial z}\right) = -\frac{\partial p}{\partial z} + \mu\left(\frac{\partial^2 v_z}{\partial r^2} + \frac{1}{r}\frac{\partial v_z}{\partial r} + \frac{\partial^2 v_z}{\partial z^2}\right). \quad (1.46)$$

We can further simplify these equations if we introduce the stream function ψ defined by

$$v_z = \frac{1}{r}\frac{\partial\psi}{\partial r}, \quad v_r = -\frac{1}{r}\frac{\partial\psi}{\partial z}.$$

With this form of ψ, Eqs. 1.45 and 1.46 can be combined to yield the following equation in ψ

$$\frac{\partial}{\partial t}(\nabla_1^2\psi) - \frac{1}{r}\frac{\partial(\psi, \nabla_1^2\psi)}{\partial(r, z)} - \frac{2}{r^2}\frac{\partial\psi}{\partial z}\nabla_1^2\psi = \nu\nabla_1^4\psi \quad (1.48)$$

where

$$\nabla_1^2 \equiv \frac{\partial^2}{\partial r^2} - \frac{1}{r}\frac{\partial}{\partial r} + \frac{\partial^2}{\partial z^2}, \quad \nabla_1^4 \equiv \nabla_1^2(\nabla_1^2)$$

$$\text{and} \quad \frac{\partial(\psi, \nabla_1^2\psi)}{\partial(r, z)} = \frac{\partial\psi}{\partial r}\frac{\partial(\nabla_1^2\psi)}{\partial z} - \frac{\partial\psi}{\partial z}\frac{\partial(\nabla_1^2\psi)}{\partial r}. \quad (1.49)$$

1.4 Initial and Boundary Conditions

The above equations can be solved subject to initial and boundary conditions. Initial conditions give the motion of the fluid at time $t = 0$ while boundary conditions are prescribed on the surfaces with which the fluid may be in contact. Contact is preserved between the fluid and the rigid surfaces and at a solid-fluid interface the fluid must not penetrate the solid if it is impermeable to the fluid. Therefore the normal velocity of the contact surface and that of the fluid are the same, ie., the relative velocity component of the fluid normal to the solid surface must vanish. In the case of a fixed rigid boundary, the boundary conditions are provided by the no-slip condition according to which both tangential and normal components of the fluid velocity vanish at all points of the surfaces of the stationary bodies with which the fluid may be in contact. However, if the rigid boundary is in motion, the relative velocity of the fluid is tangential to the surface; this is also true for two fluids in contact along a common surface.

1.5 Dimensional Analysis in Fluid Mechanics

In fluid mechanics a great many problems may be solved using dimensional analysis. The relevant dimensionless parameters in fluid mechanical situations may be combined into independent dimensionless groups characterizing the flow. These dimensionless parameters, or π's as they are called, may be obtained by the method of dimensional analysis. Dimensional analysis, or the *Buckingham's π-Theorem* as it is called (Buckingham [10]), is a method of finding the relevant dimensionless parameters without knowing the relevant differential equations. In fact, the essence of the theorem states that if a physical process is governed by a dimensionally homogeneous relation involving n-dimensional parameters then there exists an equivalent relation involving a smaller number, $(n - k)$, of dimensionless parameters, called π-variables. The reduction, k, is usually equal to but never more than, the number of fundamental dimensions involved in the physical process. However, the method requires a knowledge of all the relevant variables in any physical problem. For example, as explained earlier, for the flow of viscous incompressible fluid only the continuity equations and the Navier-Stokes equations are necessary to describe the flow. There are four independent variables u, v, w and p — the three velocity components and the pressure. In addition, there are the fluid properties, density ρ, viscosity μ, and the gravitational potential. In fact, in the Navier-Stokes equations (1.26)-(1.28), the terms on the left hand sides represent the *inertial forces* while the three terms on the right-hand side of each equation represent respectively the *body forces, pressure forces and viscous forces.* Let us put these equations in dimensionless form. Choose a characteristic velocity V and a characteristic length L. For example, if we were to consider the flow in a tube, V could be the mean flow speed and L, the tube diameter. On the other hand, if we were to consider the external flow over a cylinder, L could be the diameter and V the free stream velocity. Having chosen these characteristic quantities, we introduce the following dimensionless variables:

$$x' = x/L, \quad y' = y/L, \quad z' = z/L,$$

$$u' = u/V, \quad v' = v/V, \quad w' = w/V, \tag{1.50}$$

$$p' = p/\rho V^2, \quad t' = Vt/L$$

and the parameter

$$R_e = VL\rho/\mu. \tag{1.51}$$

In terms of these variables, Eqs. 1.26-1.28 in the absence of body forces, can be written in dimensionless or normalized form:

$$\frac{\partial u'}{\partial t'} + u'\frac{\partial u'}{\partial x'} + v'\frac{\partial u'}{\partial y'} + w'\frac{\partial u'}{\partial z'} = -\frac{\partial p'}{\partial x'} + \frac{1}{R_e}\left(\frac{\partial^2 u'}{\partial x'^2} + \frac{\partial^2 u'}{\partial y'^2} + \frac{\partial^2 u'}{\partial z'^2}\right) \tag{1.52}$$

and two similar equations in v' and w'. The equation of continuity (1.11) can also be put in dimensionless form

$$\frac{\partial u'}{\partial x'} + \frac{\partial v'}{\partial y'} + \frac{\partial w'}{\partial z'} = 0. \qquad (1.53)$$

The parameter R_e defined in Eq. 1.51 is known as the *Reynolds number* which is a dimensionless quantity, probably the most important dimensionless parameter in fluid mechanics. It is clear that only one physical parameter, the Reynolds number R_e, enters into the basic equations of fluid flow. Thus, the flows over two geometrically similar bodies (same shape, but different in size) will be identical if the Reynolds numbers for the two bodies are the same.

Consider now the physical significance of the Reynolds number. It will be shown that the Reynolds number is a measure of the ratio of inertial forces to the viscous forces. In a flow the inertial forces, in terms of characteristic velocity and characteristic length are of the order $\rho V^2/L$ and the viscous forces are of the order $\mu V/L^2$. The ratio of these two forces is

$$\frac{\text{Inertial forces}}{\text{Viscous forces}} = \frac{\rho V L}{\mu} = \text{Reynolds number}. \qquad (1.54)$$

Clearly, when the Reynolds number is small, $R_e \ll 1$, the viscous forces dominate. A small Reynolds number will occur when the viscosity is large or when either the characteristic length or characteristic velocity is small. A flow with small Reynolds number is sometimes called *slow* motion. On the other hand, when the Reynolds number is large, $R_e \gg 1$, the inertial forces are large compared to viscous forces, and the inertial forces tend to diffuse the fluid particles, causing rapid mixing in the fluid, which is a characteristic of *turbulence*.

In a straight pipe, turbulence occurs when R_e exceeds a critical value of about 2000. In the case of blood flow, laminar flow continues to occur at Reynolds numbers as high as 10,000. In fact, in many biological situations high Reynolds number laminar flow is more common than turbulent flow.

Now that we have introduced the properties of fluids and have some idea of how a fluid behaves under certain circumstances, we can put our study of fluid mechanics into perspective by developing its application in biological systems. This then will be the essence of the following chapters.

Chapter 2

CIRCULATORY BIOFLUID MECHANICS

2.1 General Indroduction

A unicellular organism can simply absorb nutrients, required to maintain its life, from its immediate environment, and just as easily excrete its waste products into the environment. Life is maintained so long as the environment is large enough to provide ample supply of nutrients, and is not polluted.

However, in the human body, which is a multi-cellular body, each cell acts as if it were a unicellular organism — it takes nutrients from its immediate environment and excretes waste-products into the same environment. This immediate environment is enclosed within the body. It is not large, and hence there is a need for a system which can transport materials (both nutrients and waste products) around the body. Such a system does exist and is termed the *circulatory system.*

The circulatory system of a living body functions as a means of transporting and distributing essential substances (eg. oxygen, hormones) to the tissues and removing by-products (eg. carbon dioxide, water) of metabolism. Apart from serving as an important transport system, the circulatory system is also essential in such homeostatic mechanism as regulation of body temperature, humoral communication throughout the body, and adjustment of oxygen and nutrient supply in different physiological situations.

Hence there must be adequate circulation at all times to the important organs of the body — brain, heart and lungs. Should circulation fail or malfunction, various diseases and even death could occur. Thus, the importance of the circulatory system

17

cannot be overemphasized.

In this chapter, a description of the circulatory system is given and several diseases related to abnormalities, malfunction or simply failure of the circulatory system are discussed.

2.2 The Circulatory System

2.2.1 *Introduction*

The circulatory system is made up of the heart and blood vessels. It performs one main function — to transport. This system is responsible for delivering oxygen and nutrients needed for metabolic processes to the tissues, carrying waste products from cellular metabolism to the kidneys and other excretory organs for elimination, and circulating electrolytes and hormones needed to regulate body function. It also regulates the body temperature by moving core heat to the periphery, where it can be dissipated into the external environment. In addition, it plays the vital role of transporting various immune substances that contribute to the body's defence mechanisms.

In other words, the circulatory system basically consists of a central pump (the heart) and the peripheral blood vessels (arteries, veins and capillaries). The transport medium, blood, is circulated throughout the body in a network as depicted in Fig. 2.1. The peripheral blood vessels form a sequential continuum insofar as blood flow is concerned, ie., blood flows successively from one type of the vessel into another. The circuit is complete as can be seen from Fig. 2.1.

2.2.2 *Systemic and Pulmonary Circulations*

The circulatory system can be divided into two sections: the systemic and the pulmonary circulations (see Fig. 2.2).

The systemic circulation supplies blood to all of the body's tissues except the lungs. As it needs to pump blood to distant parts of the body, often against gravity, the systemic circulation functions as a high-pressure system with arterial pressure up to 100 mm Hg. The lungs are supplied by the pulmonary circulation, which provides a gas exchange function. It is located in the chest in close proximity to the heart, which propels blood through it. Thus, the pulmonary circulation functions as a low-pressure system.

In Table 2.1, the first horizontal row shows the mean pressure (mm Hg) in the various blood vessels for the systemic circulation while the second horizontal row shows that for the pulmonary circulation.

The arteries carry blood pumped by the heart under pressure to the tissues and various organs of the body, supplying them with the essential substances via smaller vessels called arterioles and capillaries. After useful substances are absorbed into the tissues, blood is returned to the heart by the low-pressure smaller veins (venules) and the larger veins (vena cava). The structural properties of arteries and veins are explained in Section 3.5.

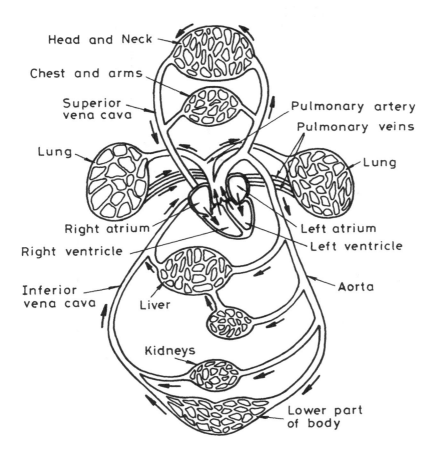

Fig. 2.1: Circulation of blood.

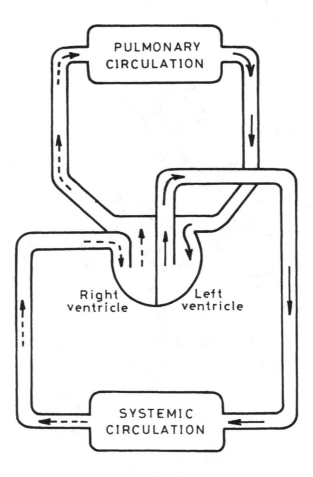

Fig. 2.2: The systemic and pulmonary circulation are arranged in series. Arrows indicate direction of blood flow (full arrows : oxygenated blood, dotted arrows : deoxygenated blood).

Table 2.1: Mean pressures (mm Hg) in blood vessels

Aorta	Arteries	Arterioles	Capillaries	Venules	Veins	Vena Cava	RA	RV
100	90 →	75 →	45 →	25 →	10 →	5 →	5 →	25
↑								↓
LV	LA	Veins	Capillaries	Arterioles		Arteries		Pulmonary artery
100	← 5	← 5	← 10	← 15	←	20	←	25

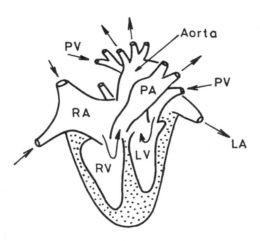

Fig. 2.3: The heart.

In essence, the efficient functioning of the circulatory system is essential to the well-being of the human body. For the system to function smoothly, a smooth flow through the distributory system is necessary.

2.2.3 *The Circulation in the Heart*

The distributory system comprises of arteries and arterioles responsible for sending blood to the various organs of the body. Fine capillaries make up the diffusing system and act as an exchange system. The collecting system consists of veins which collect blood depleted of oxygen and full of products of metabolic processes of the cardiovascular system.

Each division of the circulation includes a pump, a distributory system, a diffusing system and a collection system. The heart is the pump that propels blood through both circulations. It is divided into a right heart, which delivers blood to the pulmonary circulation and a left heart, which delivers blood to the systemic circulation (see Fig. 2.3). Each side of the heart is further divided into two chambers, an atrium and a ventricle. The ventricles are the main pumping chambers of the heart. The atria act as collection chambers for various blood returning to the heart and as auxiliary pumps for the ventricles. Thus, the heart is a four-chambered structure that generates the force to make the blood flow. There has not been a pump or engine devised that can even equal the long-term performance of the heart. Consider, for instance, an average man with a heart-beat rate of 70 beats per minute lived for 70 years. His heart would have completed about 2.6×10^{10} beats before it stopped.

As mentioned earlier, the heart actually consists of two parts — the right heart and the left heart, connected in series : the output of one part becomes the input of the other. Blood that returns from the systemic circulation (see Fig. 2.2) which supplies relevant and essential substances to the body, enters the right side of the heart. The blood is, under normal circumstances, deoxygenated (having given up its oxygen content to various parts of the body) when it enters the right heart. It is then pumped through the pulmonary circulation (the lungs) where it "collects" oxygen from and releases carbon dioxide to the air sacs of the lungs. Now with a fresh supply of oxygen, it returns to the left heart, which upon the next pump, sends blood into the systemic circulation again. Unidirectional flow of blood within the circulatory system is achieved by means of four valves strategically situated in the inflow and outflow tracts of each ventricle; this is also assisted functionally by suitable pressure gradients from arterial to venous parts of the circuit.

The "pumping" action of the heart is a two-phase process : the contraction phase when blood is forced out of the heart is known as *systole*; the relaxation, or filling phase when blood is received into the heart is termed *diastole*. During systole, the pressures in the aorta (main artery from left heart) and the pulmonary artery rise to 120 mm Hg and 25 mm Hg respectively. At the end of the systole, the valves close, and there is an abrupt drop in pressure inside the heart (intraventricular pressure). In order that blood flow is not abruptly disturbed or interfered by this phasic change, the arteries and other vessels must have some characteristic structure.

Different vessels of the circulatory system have quite different structural, and hence physiological properties. In this section, we shall concern ourselves mainly with structure and morphology of arteries and arterioles, making only brief comments on capillaries, veins and other vessels.

The basic morphologic arrangement of the wall structure of a typical artery is shown in Fig. 2.4. Note that there are in fact two layers of elastic membranes on the wall of the artery. The innermost layer of the artery is the endothelium. Depending on the composition of the constituents, the relative diameter and the distance from the heart, the arteries can be classified as elastic arteries, muscular arteries or arterioles.

Elastic arteries are large and have within their walls prominent series of sheets composed of a sclerprotein, elastin. The wall of a large elastic artery, eg. the aorta, is thick. In proportion to the cross-sectional area of the lumen, (the "space" enclosed by the arterial wall) however, the aortic wall is relatively thinner than the wall of the smaller muscular arteries. The walls of some large thick-walled arteries have their own blood supplies.

Muscular arteries are blood vessels intermediate between the large elastic arteries and the smaller arterioles. Being the most numerous arterial structure to be found in the body, muscular arteries form the major distributing arterial system. The walls of this class of artery include a thick layer of smooth muscle fibres, with collagenous tissues. Elastic connective tissues are also present. These arterial components are continuous with surrounding loose connective tissues, and thus their movements permit the vessel to change its diameter readily. In mathematical or physical terms, the vessels are not rigid but elastic. Longitudinally, however, the arteries are under a state of tension and this property is evident from the fact that the vessels retract when severed (Fung [19]).

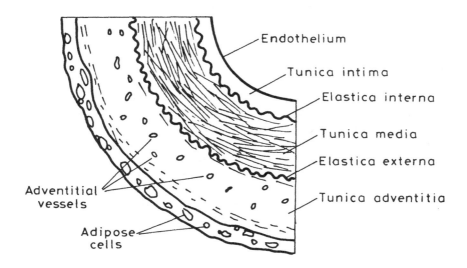

Fig. 2.4: Cross-section of a typical artery to illustrate the several layers of tissue that are present in such vessels.

The arterioles are the smallest arteries with diameter 0.3 mm or less. Only larger arteries have an internal elastic membrane, and although external elastic membranes are also found in smaller arteries, they are absent in arterioles. However, loose connective fibres including elastic and collagenous fibres are present in arterioles.

The anatomic transition from one type of artery to the next may be abrupt or progressive but a smooth flow of blood is generally maintained in a normal body. One important point to note here is the fact that the entire cardiovascular system has an inner endothelial lining, which makes it possible for it to be a smooth and continuous, uninterrupted channel in which for blood to flow.

However, the contractions of the heart are intermittent, or pulsatile and so blood is ejected in spurts and not a smooth stream. If the walls of the arteries and arterioles were rigid, then the flow of blood in the capillaries would likewise be pulsatile instead of steady. Since the exchange of substances between blood and interstitial fluid takes place through the thin walls of the capillaries, it is necessary to have a smooth, continuous flow in the capillaries. Fortunately, the large arteries near the heart are normally elastic and hence distensible. Consequently, they damp the pulsatile systolic output of the ventricles. During contraction, the ejection of blood distends the aorta and its large arterial branches. With the closure of the aortic valves, the elastic arteries recoil, thereby sustaining the pressure head and making the blood flow to the other peripheral vessels steadier than it would have been otherwise. The artery thus acts as an ancilliary pumping mechanism to the heart.

Mathematically, the elastic properties of the arteries provide an "energy-storing mechanism" — so that only a fraction of the work performed during systolic phase is expended in moving the ejected blood along the peripheral vessels, and the rest is used in dilating or distending the elastic walls. Energy is hence stored as potential energy not unlike a compressed spring. This potential energy is then expended during diastole (see Fig. 2.5). Such is the very characteristic of arteries which make them so vital in the proper functioning of the circulatory system.

The capillaries, together with the smallest venules, are the smallest blood vessels. The endothelial cells of the capillaries are capable of contraction. Hence, regulating capillary diameter and blood flow is unnecessary in capillaries. Veins are present in higher abundance than the arteries, their walls are thinner and their diameters are larger. Veins are much less elastic and tend to collapse more readily. Muscular and elastic properties are poorly, if at all, developed in veins. The basic morphological properties of blood vessels are summarised in Fig. 2.6.

2.3 Diseases Related to Circulation

Let us examine briefly some of the diseases related to the circulatory system. Vascular diseases are undoubtedly important. They weaken vessels and make them vulnerable to rupture. Such diseases also narrow and occlude the vascular lumens, thereby cutting down the blood supply to the tissues and organs. Common arterial

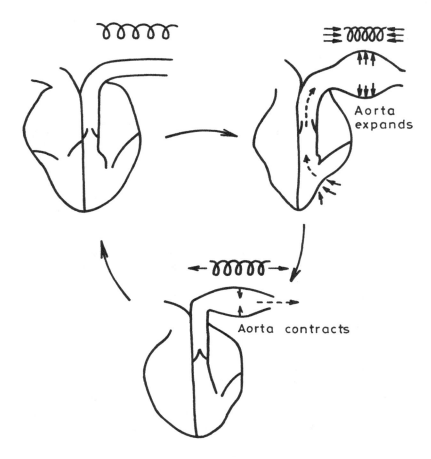

Fig. 2.5: Stages of contraction and dilation of aortic walls, illustrating the "energy-storing" mechanism with analogy to that of a spring.

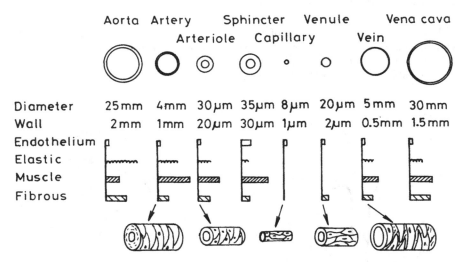

Fig. 2.6: Summary of various properties of different blood vessels.

diseases include Arteriosclerosis, Vasculitis, Raynaud's Disease, and Aneurysms. Venous diseases like Varicose Veins and venous thromboses are also common.

Arteriosclerosis is the generic term for three patterns of vascular diseases, all of which cause thickening and inelasticity of arteries. The dominant pattern is Atherosclerosis. It is the most common ailment of the Western populations. The other patterns are Mönckeberg's medial calcific sclerosis affecting the muscular arteries, and arteriolosclerosis which involves smaller arteries and arterioles.

As a person gets older, their arteries may have large amounts of cholesterol deposits, or calcifications may occur in the media of muscular arteries (in individuals over 50 years of age). This changes the structure of the arteries, and the most important changes are the lowering of elasticity of arterial walls, and the thickening of walls (hence reducing lumens). We have seen the importance of the elastic property of arteries in maintaining a smooth blood flow. With inelastic arteries and reduced lumens, coronary diseases like myocardial infarction ("heart attack") may occur. "Infarction" is a general term used to describe the situation when death of tissues occurs as a result of oxygen deficiency. A particular organ may fail depending on the degree of damage. In myocardial infarction, the organ is the heart.

Vasculitis is characterized by the inflammation of an artery (or vein) caused by the spread of a contiguous inflammatory process, such as a bacterial infection. Abnormal dilatations of arteries or veins are called Aneurysm — occurring more frequently in arteries, especially in the aorta. They are developed whenever there is a marked weakening of the wall of a vessel. Any vessel may be affected by a wide variety of disorders, including congenital defects, local infections (mycotic aneurysms), trauma or systemic diseases that weaken arterial walls.

It is not the purpose of this section to dwell on the details of the various diseases, which forms too large a subject by itself. Nevertheless, one should by now realise the extensive relevance and importance of the blood vessels and blood flow in pathology. Variations from the normal characteristics of blood flow vessels or blood itself can lead to some serious clinical problems. We will discuss this aspect in greater detail in the next chapter.

Chapter 3

BLOOD RHEOLOGY: PROPERTIES OF FLOWING BLOOD

3.1 General Introduction

Having discussed the structural properties of the arteries, we now turn our attention to the transport medium, namely, the blood.

Physiologically, amongst other functions, blood serves as the major bulk transportation and distribution system for materials within the body — transport of respiratory gases, nutrients, waste products of metabolism, etc. Because blood is a suspension of particles in an aqueous solution, it is not a homogeneous fluid and has some unusual properties.

The previous chapter confirms the importance of a knowledge of blood vessels and blood flow characteristics, as it may help in detecting, if not designing a treatment for, the various diseases. For instance, the disease, Diabetes Mellitus is related to the unusual viscous properties of blood. It has been studied and found that the viscosity of diabetic blood may be raised or lowered depending on the shear rate of blood. Studies have also been made to determine the correlation between viscosity factors of blood and chronic anxiety states. The patients, apart from having anxiety disorders were supposed to be in normal health and yet blood viscosity was higher in them.

3.2 Blood Composition

Firstly, let us look at the constituents of blood.

Table 3.1: Blood constituents (5×10^6 particles/mm^3).

Cell Elements	Relative Proportions
Red cells (erythrocytes)	600
White cells (leucocytes)	1
Platelets (thrombocytes)	30
Plasma	Weight Fraction
Water	0.91
Proteins	0.07
Inorganic solutes	0.01
Other organic substances	0.01

Table 3.2: Normal range of values for the cellular elements concentration in the human blood.

	Average cell concentration cells/ml	Approximate normal range cells/ml	Characteristic length μm	Percentage of total volume cells %
Total W.B.C.	9000	4000-11000	10	5
Neutrophils	5400	3000-6000		
Eosinophils	270	150-300		
Basophils	60	0-100		
Lymphocytes	2730	1500-4000		
Monocytes	540	300-600		
Total R.B.C.			1.9×8.0	91
Female	4.8×10^6			
Male	5.4×10^6			
Platelets	3000000	150000-400000	3	4

Blood, a complex body fluid, is a liquid tissue consisting of several types of formed elements (or cells) suspended in an aqueous fluid matrix (the plasma). The plasma itself contains a complex spectrum of organic molecules. It is made up of Red Cells (RBC or erythrocytes), White Cells (WBC or leucocytes) and Platelets (thrombocytes). Note that RBC constitute about 95%. Plasma is an aqueous solution of electrolytes and organic substances, mainly proteins. These proteins include fibrinogen, globulins, albumin, beta lipoprotein and lipalbumin. The first three make up approximately 5%, 45% and 50% of plasma protein respectively. The latter two are present in very small proportions. Fibrinogen, though not present in high concentration, is the largest of the proteins while albumin is the smallest. Tables 3.1 and 3.2 show the relative proportions of cell elements and the concentrations of plasma.

The WBC play a major role in fighting diseases. WBC are relatively fewer in number compared to the RBC, and so their contribution to the circulatory characteristic parameters is insignificant in general, though in some pathological cases, this may not be true. The platelets are very numerous, but they are also very small in size. Because of their small size, they are usually neglected or considered unimportant where blood flow discussion is concerned.

The RBC consist of a very flexible membrane enclosing concentrated solution haemoglobin. The viscosity of haemoglobin is 5 times that of blood, and this makes it possible for RBC to deform. This property enables RBC, of average diameter 8 μm not only to pass through capillaries (5 μm in diameter), but also through the endothelial wall. RBC play a major role in determining the mechanical properties of blood. It is hence important to know the volume concentration of RBC. The quantity, haematocrit is used to represent this volume concentration, being defined simply as a percentage volume concentration. It ranges from 42 to 45 in normal blood composition.

The specific gravity of red cells is about 1.06, while that of plasma is 1.03, consequently if blood is allowed to stand in a container, the red cells will settle out of suspension. They do so at a definite rate called the *Erythrocyte Sedimentation Rate* or ESR for short. The mathematical theory of this process is very simple and can be studied on the basis of the theory of fluid mechanics. In illness, the ESR always increases abruptly. This is due to the formation of red cell aggregation.

3.3 Structure of Blood

Blood cells form a continuous structure when at rest (see Fig. 3.1). Sedimentation occurs when the structure breaks under its own weight and when a finite stress is applied to the whole blood, the structure breaks. The stress needed to cause this breakage is called the yield stress. After this breakage, the structure becomes a suspension of a cluster of cell aggregates in plasma. These aggregates are in turn formed from smaller units of cells called rouleaux [6].

There exists a dynamic equilibrium between the size of aggregates or rouleaux and the stress applied. Under sufficiently high stress, aggregates and rouleaux are reduced to individual cells, which will not be reduced any further by further increases in stress. Thus, the shear stress-shear rate relationship becomes a straight line at sufficiently high shear stresses (see Fig. 3.2).

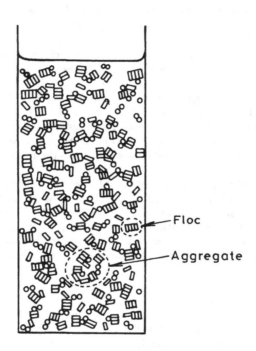

Fig. 3.1: A suspension of cells at rest is a continuous structure formed from groups of cell aggregates made up of units of cells called rouleaux or flocs.

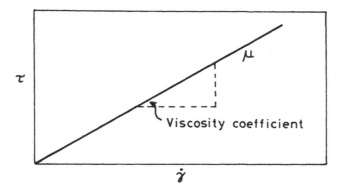

Fig. 3.2: Dynamic equilibrium between cell aggregate and shear rate normal plasma; hematocrit $\simeq 45\%$).

In fact, the shear stress-shear rate relationship of plasma is normally considered linear as shown in Fig. 3.3. This is, of course, the property of a Newtonian fluid. Thus, plasma is considered to be Newtonian. However, there are several studies which describe plasma as non-Newtonian. It can thus be concluded that plasma may have transitory non-Newtonian character at low shear rates.

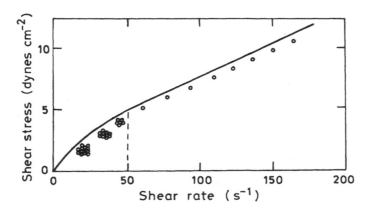

Fig. 3.3: The relationship between shear stress and shear strain rate for blood plasma.

3.4 Flow Properties of Blood

3.4.1 *Viscosity of Blood*

Blood properties are dominated by the properties of RBC as about 95% of blood cells are RBC (except in cases like leukaemia) (see Table 3.1). A human red cell is a disc-shaped element having a mean diameter of 7.2 μm and a thickness of about 2.2 μm at the thickest part, that is, near the circumference, and about 1 μm at the center (see Fig. 3.4).

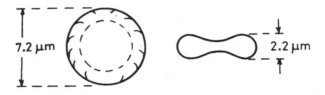

Fig. 3.4: Diagram showing the dimensions of the red cell.

It has an average area of 120 μm^2 and a volume of 85 μm^3. The biconcave shape serves to increase the surface area available for diffusion of oxygen and allows the cell to change in volume and shape without rupturing its membrane.

The red cell is bounded by a membrane made of protein in association with lipid and steroid materials, a semi-permeable membrane. The most important element in the red cell is the haemoglobin, which transports oxygen to the tissues.

The volume fraction of red cells in blood is known as the hematocrit, H. However, this value is normally a little greater than the true volume fraction, ϕ, as a small volume of plasma is trapped between the cells.

Blood is neither homogeneous nor Newtonian. Plasma, when isolated, may be Newtonian with a viscosity of about 1.2 times that of water. However, the viscosity for whole blood is found to depend on the shear rate and is measured as apparent viscosity (see Fig. 3.3). It is thus a non-Newtonian fluid.

For values of ϕ up to 0.05 a constant shear stress-shear strain rate relationship typical of a Newtonian fluid is observed. The viscosity of blood for $\phi < 0.05$ is expressed with reasonable accuracy by Einstein's equation for spheres in suspension (Jeffery [31])

$$\mu_s = \mu_p \left(\frac{1}{1 - \alpha\phi} \right) \tag{3.1}$$

where μ_s = suspension viscosity
μ_p = plasma viscosity
α = shape factor (2.5 for spheres)
ϕ = cell volume fraction.

In theory, the constant α should vary with the shape of the particles, but a value of 2.5 appears to apply to red cells when $\phi < 0.05$.

However, Eq. 3.1 is only good under the assumption that the particles act individually and do not interact significantly. As ϕ increases from 0.05, this assumption is not met as interaction between particles becomes more significant. The properties associated with interaction, for example, cell elasticity and aggregation, now influence viscosity, which departs from Eq. 3.1.

Later, the use of Eq. 3.1 with a variable α is proposed, where

$$\alpha = 0.076 \exp \left[2.49\phi + \frac{1107}{T} e^{(-1.69\phi)} \right] \tag{3.2}$$

and T = absolute temperature ($^\circ K$).

This is found to agree with measurements of apparent viscosity of red cell suspensions for ϕ up to 0.6. Figure 3.5 shows a plot of α against ϕ for $T = 25^\circ$ and 37°.

3.4.2 Yield Stress of Blood

A number of non-linear expressions have been suggested as a means to characterize blood viscosity and yield stress of blood. For example, Casson [11] suggested an expression

$$\sqrt{\tau} = K\sqrt{\gamma} + \sqrt{c} \tag{3.3}$$

whereas Herschel-Bulkley suggested

$$\tau = b\gamma^s + c \tag{3.4}$$

where K = Casson viscosity,
c = yield stress of blood, and
b, s = constants.

As mentioned before, blood requires a finite stress to disrupt the aggregates before it begins to flow. For hematocrits $H > 5.8\%$, this yield stress is found to be given by

$$\tau_0^{\frac{1}{3}} = \frac{A(H - H_m)}{100} \tag{3.5}$$

where $A = (0.008 \pm 0.002 \, \text{dyne/cm}^2)^{\frac{1}{3}}$

$\quad\quad H$ = normal hematocrit

$\quad\quad H_m$ = hematocrit below which there is no yield stress.

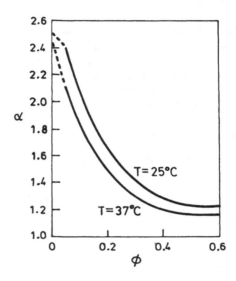

Fig. 3.5: Plot of α against ϕ for ϕ up to 0.6.

Taking $H = 45\%$ and $H_m = 5\%$, the yield stress of normal human blood works out to be between 0.01 and 0.06 dyne/cm^2.

Other than the yield stress τ_0, shear stress τ and apparent viscosity μ are also dependent on the hematocrit. On top of this, the apparent viscosity is observed to depend on capillary radius for capillaries of diameter less than 300 μm. This phenomenon is known as the Fahraeus-Lindqvist effect.

The yield stress is determined from the cell to cell contact (see Fig. 3.1). A number of ingenious methods have been used for the measurement of yield stress. However, there is an inherent difficulty in the measurement of yield stress. The measuring instruments interact with the red cell structure and little is known about this interaction. Hence, it is not apparent whether measurements of blood yield stress are in fact measurements of the structure of blood or the structure-instrument interface. We will discuss blood viscosity again in the next chapter after we discuss models of blood flow.

3.5 Blood Vessel Structure

Except for the capillaries, all blood vessels have walls composed of three layers separated by an elastic lamina (see Fig. 3.6). The tunica externa, or tunica adventitia, is the outermost covering of the vessel. This layer is composed of fibrous and connective tissue that serves to support the vessel. This layer is separated by the external elastic lamina from the middle layer, or the tunica media, which is largely a smooth muscle that constricts and relaxes in order to control the diameter of the vessel. The tunica intima, or inner layer, has an elastic layer called the internal elastic lamina, that joins the media and a thin layer of epithelial cells that lie adjacent to the blood. The epithelial layer provides a smooth and slippery inner surface for the vessel. This smooth inner lining, as long as it remains intact, prevents blood clotting.

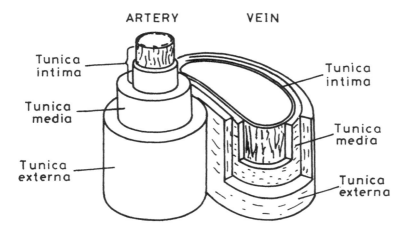

Fig. 3.6: Medium-sized artery and veins showing the relative thickness of the three layers.

3.5.1 *Arteries and Arterioles*

The layers of the vessel vary with the vessel's function. Arteries are thick-walled vessels with large amounts of elastic fibres. The elasticity of these vessels allows them to stretch during cardiac systole and recoil during diastole. The arterioles which are predominantly smooth muscles, serve as resistance vessels for the circulatory system. They constrict or relax as needed to maintain blood pressure.

3.5.2 *Veins and Venules*

The veins are thin-walled, distensible, and collapsible vessels. The structure of the veins allows these vessels to act as a reservoir, or blood storage system.

3.5.3 *Capillaries*

The capillaries are microscopic single-thickness vessels that connect the arterial and venous segments of the circulatory system. Exchange of gases, nutrients, and waste materials occurs through the thin permeable walls of the capillaries.

The approximate dimensions of these human blood vessels can be found in Table 3.3.

Table 3.3: Approximate dimensions of human blood vessels.

Vessel	Diameter (cm)	Length (cm)	Wall Thickness (cm)	Average Velocity (cm/s)
Capillaries	0.0008	0.1	0.0001	0.1
Venules	0.002	0.2	0.0002	0.2
Arterioles	0.005	1	0.02	5
Arteries	0.4	50	0.1	45
Veins	0.5	2.5	0.05	1.0
Aorta	2.5	50	0.2	48
Vena Cava	3.0	50	0.15	38

3.6 Diseases Related to Obstruction of Blood Flow

There are various mechanisms of vessel obstruction. These include (i) thrombus formation, (ii) emboli, (iii) compression, (iv) vasospasm, and (v) structural defects in the vessel (see Fig. 3.7).

3.6.1 *Thrombus Formation*

A thrombus is a blood clot. It can develop in either the arterial or the venous system and obstruct flow. It may cause acute arterial occlusion and, in extremity, will result in sudden onset of acute pain with numbness, tingling, weakness, pallor, and coldness. These changes are rapidly followed by cyanosis, mottling, and loss of sensory, reflex, and motor function. Thromboangiitis obliterans is an inflammatory arterial disorder which causes thrombus formation that affects medium-sized arteries.

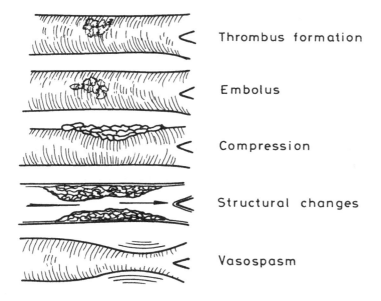

Fig. 3.7: Conditions that cause disruption of blood flow.

3.6.2 *Embolus*

An embolus is a foreign mass that is transported in the bloodstream. Although it moves freely in the larger blood vessels, it becomes lodged and obstructs flow once it reaches a smaller vessel. An embolus can be a dislodged thrombus or it can consist of air, fat, tumor cells or other materials. Arterial emboli commonly have their origin in the heart itself and can travel to the brain, spleen, kidney or vessels and the lower extremity before they become lodged and obstruct flow. Acute arterial occlusion normally results from an embolus. Its effects are described above. Emboli tend to lodge in bifurcations of major arteries, including the aorta and iliac arteries.

3.6.3 *Compression*

The lumen of a blood vessel can be occluded by external forces. The external pressure of a tourniquet, for example, is intended to compress blood vessels and interrupt blood flow. Similarly, casts and circular dressings predispose to vessel compression, particularly when swelling occurs after these devices have been applied. Tumors may encroach on blood vessels as they grow. Blood vessels may be compressed between bony structures such as the sacrum and the supporting surfaces of a chair or bed.

3.6.4 *Vasospasms*

Vasospasm may result from locally or neurally mediated reflexes. Exposure to cold causes severe vasoconstriction in many of the superficial blood vessels. In certain disease states, vasospasm from exposure to cold or other stimuli is excessive and may lead to ischemia and tissue injury. Intense vasospasm of the arteries and arterioles in the fingers and toes causes Raynaud's syndrome, a type of peripheral arterial disease.

3.6.5 *Structural Defects*

Structural changes in blood vessels can take many forms. As explained in Section 2.3, arteriosclerosis causes rigidity and narrowing of the arteries. One form of arteriosclerosis, Mönckeberg's medial sclerosis affects the media of medium-sized arteries and is characterized by thickening of the arterioles. The most common form of arteriosclerosis — atherosclerosis, affects both the large and medium-sized arteries. Complications of atherosclerosis include heart attack, aneurysms, stroke, and peripheral vascular disease.

Chapter 4

MODELS OF BIOFLUID FLOWS

4.1 Flows in Pipes and Ducts

4.1.1 *Introduction*

As mentioned earlier in Chapter 1, the flows of fluid around objects can be divided into two regions: a thin region close to the object where viscosity effects are important (Boundary Layer Flow), and an outer region where viscosity can be neglected (Potential Flow). Although there is no precise transition line between these two regions, boundary layer is usually defined to be that region where the fluid velocity parallel to the surface is less than 99% of the free stream velocity of the potential flow. The thickness of the boundary layer, δ, grows along a surface over which the fluid is flowing (see Fig. 4.1). Within the boundary layer, the flow is laminar, but as the thickness of the layer grows larger a transition region appears and the flow in the boundary layer may become turbulent. In fact, if the boundary surface is long enough, the laminar-transition-turbulence sequence occurs in all fluid flows irrespective of the nature of the free stream as to whether it is laminar or turbulent. The Reynolds number for the transition from laminar to turbulent flow is approximately 2300. However, for special conditions transition may occur at Reynolds number as high as 40,000.

41

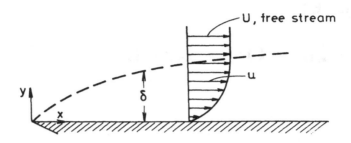

Fig. 4.1: Boundary layer flow.

4.1.2 *Developing and Fully Developed Flow*

We shall consider here flow in closed conduits. If a conduit has a circular cross-section of constant area, it is called a pipe. Conduits of other cross-sections but of constant area are called ducts.

The flow in a straight pipe or duct is said to be *fully developed* if the shape of the velocity profile is the same at all cross-sections. Here the velocity varies with distance across the pipe but not along it and there is no well defined boundary layer. The flow is one directional and there is no velocity perpendicular to the axis of the pipe. We will encounter fully developed flow in the next section, while discussing Poiseuille's flow.

One may think that fully developed flow occurs only under special circumstances. However, many pipe flows are fully developed.

Let us consider the flow from a large reservoir into a pipe (see Fig. 4.2). Consider the flow is laminar in the entry region of the pipe. If the junction between the pipe and the reservoir is nicely rounded, the velocity of the fluid entering the pipe at cross-section 1 will be nearly uniform. However, fluid in contact with the pipe wall will have zero velocity due to the no-slip condition. Hence, a region of slower-moving fluid near the wall is established. This is the boundary layer region which grows wider as the flow moves downstream. Obviously, the boundary layer grows with distance from the entrance until the flow becomes fully developed. The flow in the central region of the pipe is called the *core flow* or the *plug flow* and the region is called the *core region*. Since the fluid near the wall is moving more slowly, the core flow must accelerate as it moves downstream to maintain the constant mass flow at all cross-sections.

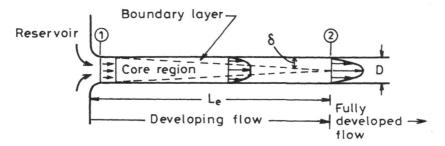

Fig. 4.2: Flow in the entry region for laminar flow.

It is to be noted that the distinction between laminar and turbulent flow in a pipe or duct applies only to the fully developed flow condition. If the fully developed pipe flow is laminar ($R_e < 2300$), the boundary layer will be laminar, but if the fully developed pipe flow is turbulent ($R_e \geq 4000$, say) then the boundary layer will be laminar near the pipe entrance, will undergo transition, and will be turbulent as it approaches the fully developed condition (see Fig. 4.3). For Reynolds number between 2300 and about 4000, the flow is unpredictable, sometimes pulsating or switching back and forth between laminar and turbulent. This type of flow is called *transitional flow*. For $R_e > 4000$, the flow is usually turbulent.

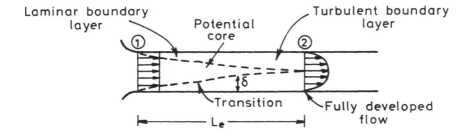

Fig. 4.3: Flow in the entry region for turbulent flow.

Clearly, the boundary layer continues to grow till it fills the entire pipe when the boundary layer thickness is equal to the tube radius. This occurs at section 2 in Figs. 4.2 and 4.3. Downstream of section 2, the flow is fully developed. The flow between sections 1 and 2 is called *developing flow* or *inlet flow* and the region is called the *entry region*. The length L_e till the flow is fully developed is called the *development length* or the *entry length*. This length for laminar flow (Fig. 4.2) as well as for turbulent flow (Fig. 4.3) for a pipe of diameter D was found by analytical and experimental investigations to be

$$L_e = 0.06 R_e D, \quad \text{for laminar flow,} \tag{4.1}$$

$$L_e = 4.4(R_e)^{\frac{1}{6}} D, \quad \text{for turbulent flow.} \tag{4.2}$$

For a laminar flow with Reynolds number $R_e = 2300$, the development length becomes

$$L_e \simeq 140D, \tag{4.3}$$

whereas for a turbulent flow with Reynolds number $R_e = 10^4$ say, the development length becomes

$$L_e \simeq 20D. \tag{4.4}$$

For pipes used in many engineering applications which are hundreds or thousands of diameters long, the flow is fully developed over most of their length.

As mentioned earlier, high Reynolds number laminar flows are more common in biological situations than in engineering situations. For example, the flow of air in the respiratory system is usually laminar, but it may become turbulent during heavy breathing or as a result of obstruction in the breathing passageway when the Reynolds number may be as high as 50,000.

The discussion in this section has been confined on flow entering a pipe from a large reservoir. Any disturbance such as placing a valve in the pipe, or changing the pipe diameter will cause the flow to depart from fully developed conditions; however, the flow returns to fully developed conditions after a while downstream.

It is to be noted that in a fully developed flow, because the shape of the velocity profile is constant, the streamlines are straight and parallel to the pipe axis, and the shear stress at the pipe wall is the same in all directions. However, it is to be remembered that the problems of blood flow are much more complicated than the problems of fluid flow in engineering situations because of unusually high Reynolds number of flow. The flow remains laminar at Reynolds numbers as high as 5000-10,000 causing the entry length (which is proportional to the Reynolds number) to be very large so that in most cases the fully developed flow is never reached because tube branchings start before this stage is attained. We will study these problems with much simplification in the following chapters.

4.2 Models of Blood Flows

4.2.1 *Introduction*

As mentioned earlier, interruption of blood flow in either the arterial or venous system interferes with the delivery of oxygen and nutrients to the tissues. Further, it is realized that unfortunately there are numerous arterial diseases which result in the occlusion of blood flow. An appropriate model will assist in the design of a more accurate method to detect these diseases. Thus, the study of models for blood flow has interested clinicians, physiologists, hemorheologists and hydrodynamicists, etc.

The present section mainly aims at finding suitable and more physically and physiologically correct models for the flow of blood. Such models may be helpful in the design of flowmeters which in turn could serve as helpful indicators where detection of diseases is concerned. Alternatively, a knowledge of plausible models can be useful when there is a need to explain perhaps some peculiar clinical or experimental results. There are existing models for blood flows, but they usually contain too many assumptions and hence restrictions on the physical properties of blood. This chapter also examines and explains the inadequacies of such models.

4.2.2 *Poiseuille's Flow*

Having justified the usefulness for a reasonable model of blood flow, we set out to describe some existing models. In this section, we discuss the Poiseuille's law model.

In 1839, the engineer Hagen [23] studied the water flow through brass tubes. Using tubes of known lengths and radii, he found that the volume flow varied as the pressure to a power of the radius and was inversely proportional to the tube length. Further experiments led him to conclude that the power should theoretically be 4. Hagen was thus the first to suggest the importance of r^4 in the flow through any tube.

Poiseuille (1799-1869) was a French physician who first investigated in a quantitative manner the flow of water through glass tubes. Poiseuille's interest was the flow of blood through the vessels of the circulatory system, but worked with water because of the difficulty at that time of preventing blood from clotting on exposure to air. He provided a more complete analysis considering the flow through a circular cylindrical tube. Using compressed air, he forced a constant water volume through capillary tubes, and discovered quite naturally that the volume of water discharged in unit time Q was proportional to the pressure head P and r^4, and inversely proportional to tube length L. This is Poiseuille's law [50,51] although it is sometimes referred to as Hagen-Poiseuille's law, and mathematically, in the most primitive form:

$$Q = K\frac{Pr^4}{L} \tag{4.5}$$

where K is a constant of proportionality.

The solution was derived by Wiedemann (1856) and Hagenbach (1860) independently. The derivation is given below:

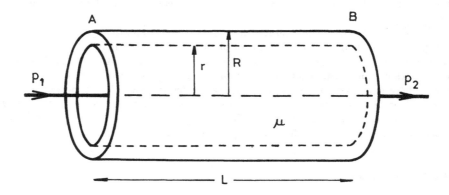

Fig. 4.4: Flow in a circular cylindrical tube.

Consider a tube of length L and radius R (Fig. 4.4) in which a viscous fluid of viscosity μ is flowing. Let p_1 and p_2 be the inflow pressure at A and the outflow pressure at B respectively, and let these be maintained constant throughout our discussion. The flow is assumed to be such that each particle of liquid moves parallel to the axis with a constant velocity. All points lying on the same circle around the axis will hence have the same velocity, so the liquid is thought of as having cylindrical laminae moving at velocities u, u being functions of their radii. This is actually Newton's hypothesis for laminar flow.

The force exerted by the difference in pressure at the ends on a cross-sectional area is:

$$F = \pi r^2(p_1 - p_2). \tag{4.6}$$

The retarding force to this motion is the viscous force F_v given by:

area × viscosity × velocity gradient.

Hence

$$F_v = (2\pi r L) \times \mu \times \frac{du}{dr}. \tag{4.7}$$

These forces are equal and opposite, hence,

$$F = -F_v$$

$$\pi r^2 (p_1 - p_2) = -2\pi r L \mu \frac{du}{dr}$$

$$r(p_1 - p_2) = -2L\mu \frac{du}{dr}$$

$$\frac{du}{dr} = \frac{-r(p_1 - p_2)}{2L\mu}. \tag{4.8}$$

Equation 4.8 is thus the expression for the velocity gradient, and on integration, yields

$$u = \frac{-r^2 (p_1 - p_2)}{4L\mu} + C \tag{4.9}$$

where C is an arbitrary constant.

If we assume that the fluid in contact with the tube wall is at rest, then $u = 0$ at $r = R$, hence,

$$C = \frac{R^2 (p_1 - p_2)}{4L\mu}. \tag{4.10}$$

Substituting Eq. 4.10 into Eq. 4.9, we get,

$$u = \frac{p_1 - p_2}{4L\mu}(R^2 - r^2). \tag{4.11}$$

Equation 4.11 is an equation for a parabola (see Fig. 4.5). Hence, the volume of the paraboloid formed by revolving this parabola around its axis is the volume of fluid Q which flows in unit time, ie.

$$Q = \int_0^R 2\pi u.r \, dr. \tag{4.12}$$

Substitution of Eq. 4.11 into Eq. 4.12 gives

$$Q = \frac{2\pi(p_1 - p_2)}{4L\mu} \int_0^R (R^2 - r^2) r \, dr;$$

$$\text{or,} \quad Q = \frac{(p_1 - p_2)\pi R^4}{8L\mu}. \tag{4.13}$$

Equation 4.13 is commonly known as Poiseuille's equation, or Poiseuille's law.

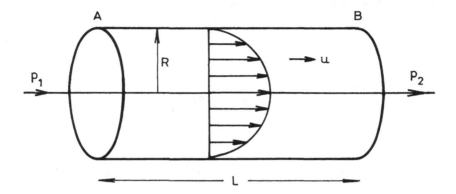

Fig. 4.5: Velocity profile using Poiseuille's law.

In the above derivation of Poiseuille's equation, which gives a plausible expression for blood flow, several conditions have to be satisfied. Amongst these conditions, the two most important are:

(1) The tube is rigid; its diameter remains constant with respect to internal pressures p_1 and p_2. We know from our previous discussion that blood vessels, in particular, arteries, are not rigid but elastic, and they distend with increasing internal pressure. This distensibility is much less in smaller vessels, in which case Poiseuille's equation may apply.

(2) The fluid is homogeneous; ie. having a constant viscosity at all rates of shear (velocity gradients). Such fluids are Newtonian fluids. We have seen that blood is not homogeneous but a suspension of cells in plasma. Hence, blood does not have a constant viscosity, and this fact is ignored by Eq. 4.13 which includes the constant μ.

At high flow rates, cells are observed to move away from the walls of the blood vessels towards the axis, leaving a plasma layer which leads to deviations from Poiseuille's equation. At low flow rates, the cells aggregate and move as clumps, so the deviations are not as pronounced. In the case where there is no marginal layer and yield stress effects are negligible, this equation is exactly correct. This is also true when the tube diameter is less than 0.05 cm.

Although it is now recognized that blood is non-Newtonian in nature, the works of Hagen and Poiseuille are not forgotten and are still used for their simplicity.

4.3 Consequence of Poiseuille's Flow

Poiseuille's law is so well established experimentally, that it is often used in order to determine the viscosity coefficient μ of viscous fluids. When blood was examined in this manner, it was found that the viscosity of blood $\mu_B = 5\mu_0$ (viscosity of water), if the diameter of the tube is relatively large. Thus at normal physiological temperature of $37°C$, $\mu_0 = 0.07P$ (P = Ppise = dyne sec cm^{-2}) and $\mu_B = 0.035P$, as determined by Poiseuille's law in large tubes.

The fact that the effective viscosity coefficient of blood according to Poiseuille's law depends on the radius of the tube in which it is measured indicates further that blood is not a Newtonian fluid, for which μ is constant.

4.4 Applications of Poiseuille's Law for the Study of Blood Flow

We have seen from Poiseuille's law,

$$Q = \frac{\pi R^4}{8\mu L}(p_1 - p_2),$$

that the flux Q is proportional to the pressure difference $(p_1 - p_2)$. However, the observed flow of blood through the veins shows a marked non-linearity (see Fig. 4.6). For small pressure differences, the flux Q is proportional to $(p_1 - p_2)$ just as in Poiseuille's law, but when the pressure difference becomes large, the flux gradually attains a maximum value and then no longer increases.

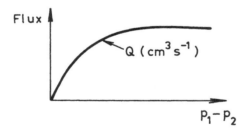

Fig. 4.6: Relationship between flux and pressure-difference in Poiseuille's flow.

A similar situation occurs in the flow through a rubber tube. This result suggests that it is important to take into account the elastic nature of the blood carrying vessels.

An artery or a vein is not a rigid tube but is rather an elastic one. Consequently, the radius of such a tube is variable. Imagine such a vessel filled with fluid at rest and surrounded by fluid (see Fig. 4.7). Let the wall thickness of the tube be h, the radius of the tube r, the exterior pressure p_0, and the interior fluid pressure be p. The change in p, namely, $(p - p_0)$, is called the *transmural pressure difference*. Under normal circumstances the thickness h is small compared to the resting radius (when exterior pressure p_0 equals the interior pressure p) of the vessel. Consequently, as a good approximation we can treat the wall as a thin elastic membrane.

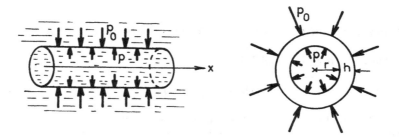

Fig. 4.7: An elastic tube filled with fluid at rest and surrounded by fluid.

Let T denote the tension per unit length of the tube and per unit thickness. Consider the equilibrium of half of such a long cylinder vessel plus the fluid contained within it. The net downward force per unit length on this half cylinder is $2Th$ and it is balanced by the net upward force per unit length which is given by

$$\int_0^\pi (p - p_0) r \sin \theta \, d\theta, \tag{4.14}$$

which equals

$$2r(p - p_0). \tag{4.15}$$

Thus we have

$$Th = r(p - p_0). \tag{4.16}$$

The tension T that develops is a property of the elastic wall in its reaction to stretch. The force per unit acting on the surface of the volume element is called stress (see Fig. 4.8), ie.,

$$\mathbf{F} = \mathbf{N} + \mathbf{H}, \tag{4.17}$$

where $\mathbf{N}$ is the normal component such as tension or pressure called the normal stress, and $\mathbf{H}$ is the tangential component called the shearing stress. In response to stress the volume element undergoes a deformation called strain. We have similarly normal strain and shearing strain.

According to Hooke's law, we have

$$T = E\frac{(r - r_0)}{r_0} \tag{4.18}$$

where $r = r_0$ in equilibrium position when the tension T is zero.

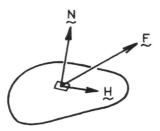

Fig. 4.8: Stress components in an elastic membrane.

Let us now consider the steady flow through an elastic tube of length L such as an artery or a vein. A diagram of the structure is given in Fig. 4.9.

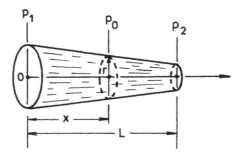

Fig. 4.9: Flow through an elastic tube.

We assume that the tube is long and the pressure is a function of x alone and that

$$p(x)|_{x=0} = p_1 \quad \text{and} \quad p(x)|_{x=L} = p_2 \tag{4.19}$$

where p_1 and p_2 are the inlet and outlet pressures respectively. The external pressure of the fluid surrounding the tube is assumed to have a constant value of p_0 (say). We have from Eq. 4.16 that

$$p(x) - p_0 = \frac{Th}{r} \tag{4.20}$$

where r is the cross-sectional radius at location x. The pressure-radius relation is obtained from Eqs. 4.18 and 4.20 giving,

$$p(x) - p_0 = \frac{Eh}{r}\frac{(r - r_0)}{r_0} = \frac{Eh}{r_0}\left(1 - \frac{r_0}{r}\right). \tag{4.21}$$

Now the flow through the tube is assumed to obey Poiseuille's law in the form

$$Q = -\frac{\pi}{8\mu}\frac{dp}{dx}r^4, \tag{4.22}$$

where

$$\frac{dp}{dx} = \frac{p_2 - p_1}{L}, \tag{4.23}$$

and r and p are related by Eq. 4.21. From Eq. 4.21 we see that r is a function of $(p - p_0)$ ie.,

$$r = f(p - p_0) = r(p - p_0)\text{(say)}. \tag{4.24}$$

Integrating Eq. 4.22 we obtain

$$\int_0^L Q\,dx = -\frac{\pi}{8\mu}\int_{p_1}^{p_2} r^4\,dp, \tag{4.25}$$

which reduces to

$$QL = \frac{\pi}{8\mu}\int_{p_2}^{p_1} r^4\,dp. \tag{4.26}$$

Making the substitution $p' = p - p_0$ and simplifying the resulting equation, gives for the volume flux

$$Q = \frac{\pi}{8\mu L}\int_{p_2-p_0}^{p_1-p_0} r^4\,dp', \tag{4.27}$$

where r is a function of p'. We hence see that Q is proportional to $1/L$ and is a function of $(p_1 - p_0)$ and $(p_2 - p_0)$. To evaluate the above integral, Eq. 4.21 is rearranged to give

$$r = \frac{r_0}{(1 - \frac{r_0}{Eh}p')}. \tag{4.28}$$

Therefore, with the use of Eq. 4.28, the definite integral appearing in Eq. 4.27 can easily be evaluated to give

$$Q = \frac{\pi r_0^3 E h}{24 \mu L} \left[\left(1 - \frac{r_0}{Eh}(p_1 - p_0) \right)^{-3} - \left(1 - \frac{r_0}{Eh}(p_2 - p_0) \right)^{-3} \right],$$ (4.29)

or

$$Q \approx \frac{\pi r_0^3 E h}{24 \mu L} \left[\left(1 + \frac{3r_0}{Eh}(p_1 - p_0) + \cdots \right) - \left(1 + \frac{3r_0}{Eh}(p_2 - p_0) + \cdots \right) \right]$$ (4.30)

which is certainly not linear. By retaining only the first order terms the above collapses to Poiseuille's law. We know that r^4 is a function of $p\prime$ (ie., $r^4(p\prime) = r^4(p - p_0)$). When $(p_1 - p_2)$ is small, r^4 is not too rapidly varying and can be taken as a constant for this period of integration and we have from Eq. 4.27,

$$Q \approx \frac{\pi}{8 \mu L} r^4 [(p_1 - p_0) - (p_2 - p_0)]$$

$$\approx \frac{\pi}{8 \mu L} r^4 (p_1 - p_2),$$ (4.31)

which like Poiseuille's law is proportional to the pressure difference $(p_1 - p_2)$.

4.5 Pulsatile Flow

Consider the flow in a circular vessel. Assuming the only non-zero component of the velocity vector is in the axial direction and is denoted by u. Using cylindrical polar coordinates, and taking the problem to be axisymmetric (ie., $u = u(r, x, t), p = p(r, x, t)$), the continuity equation reduces to

$$\frac{\partial u}{\partial x} = 0.$$ (4.32)

The Navier-Stokes equations become

$$\rho \frac{\partial u}{\partial t} = -\frac{\partial p}{\partial x} + \frac{\mu}{r} \frac{\partial}{\partial r} \left(r \frac{\partial u}{\partial r} \right),$$ (4.33)

$$0 = -\frac{\partial p}{\partial r}.$$ (4.34)

The above Eqs. 4.32 and 4.34 imply that

$$u = u(r, t) \quad \text{and} \quad p = p(x, t).$$ (4.35)

Therefore Eq. 4.33, governing the flow simplifies to

$$\frac{\mu}{r} \frac{\partial}{\partial r} \left(r \frac{\partial u}{\partial r} \right) - \rho \frac{\partial u}{\partial t} = \frac{\partial p}{\partial x}.$$ (4.36)

Observing that the left-hand side of the above equation is a function of (r, t) only and the right-hand side a function of (x, t) only, we conclude that both sides must be a function of t only.

Consider now a pulsatile sinusoidal flow with the pressure gradient and the axial velocity as

$$\frac{\partial p}{\partial x} = -Pe^{i\omega t}, \tag{4.37}$$

$$u(r, t) = U(r)e^{i\omega t} \tag{4.38}$$

where P is a constant, and $U(r)$ is the velocity profile across the tube of radius a. We assume that the flow is identical at each section along the tube so that a travelling wave solution can be neglected. It is clear that when $\omega = 0$ (the steady case), the flow becomes that of Poiseuille flow discussed earlier.

From Eqs. 4.37 and 4.38 it is clear that the real part gives the velocity for the pressure gradient $P \cos \omega t$ and the imaginary part gives the velocity for the pressure gradient $P \sin \omega t$. Upon substituting Eqs. 4.37 and 4.38 into Eq. 4.36 the following equation results:

$$\frac{d^2 U}{dr^2} + \frac{1}{r}\frac{dU}{dr} - \frac{i\omega\rho}{\mu}U = -\frac{P}{\mu}. \tag{4.39}$$

It is known that the general solution of an ordinary differential equation in the form

$$\frac{d^2 y}{dx^2} + \frac{1}{x}\frac{dy}{dx} - K^2 y = 0 \tag{4.40}$$

is

$$y = AJ_0(iKx) + BY_0(iKx), \tag{4.41}$$

which involves Bessel functions of complex argument. Thus, the solution of Eq. 4.39 is

$$U(r) = AJ_0\left(i\sqrt{(i\omega\rho/\mu)}r\right) + BY_0\left(i\sqrt{(i\omega\rho/\mu)}r\right) + \frac{P}{\omega\rho i}. \tag{4.42}$$

As U must be finite on the axis (ie., at $r = 0$), and since $Y_0(0)$ is not finite, then B has to be zero. Also, because of the no-slip condition $U(r)|_{r=a} = 0$, we have

$$AJ_0\left(i^{3/2}\sqrt{(\omega\rho/\mu)}a\right) + \frac{P}{\omega\rho i} = 0. \tag{4.43}$$

Let us now introduce a non-dimensional parameter α, known as the *Womersley parameter* given by

$$\alpha = a\sqrt{\frac{\omega\rho}{\mu}} \quad \text{or} \quad \alpha = a\sqrt{\frac{\omega}{\nu}} \tag{4.44}$$

where $\nu = \mu/\rho$ is the kinematic viscosity. From Eq. 4.43

$$A = -\frac{P}{\omega\rho i}\frac{1}{J_0(i^{3/2}\alpha)} \quad \text{or} \quad A = \frac{iP}{\omega\rho}\frac{1}{J_0(i^{3/2}\alpha)}, \tag{4.45}$$

and finally, from Eq. 4.42 we get

$$U(r) = -\frac{iP}{\omega\rho}\left(1 - \frac{J_o(i^{3/2}\alpha r/a)}{J_0(i^{3/2}\alpha)}\right).$$
(4.46)

In the limit as α approaches zero, ie., as $\omega \to 0$, the velocity profile becomes parabolic. As α tends to infinity, ie., viscosity is unimportant ($\mu \to 0$), it can be shown that

$$\frac{J_0(i^{3/2}\alpha r/a)}{J_0(i^{3/2}\alpha)} \to 0$$
(4.47)

which implies that

$$U(r) \to -\frac{iP}{\omega\rho}.$$
(4.48)

In fact, introducing the idea of a *Stokes boundary layer*, of thickness δ proportional to $1/\alpha$, it can be concluded that in this case the boundary layer thickness at the cylinder wall disappears.

It is interesting to note that the above expression is independent of viscosity, μ, and exactly 90° out of phase with P. Actually this result corresponds to the Euler equation (which only holds for $\mu = 0$, ie., inviscid fluids) namely,

$$\frac{\partial u}{\partial t} = -\frac{1}{\rho}\frac{\partial p}{\partial x}.$$
(4.49)

Using Eqs. 4.37 and 4.38 in Eq. 4.49 we easily see that

$$U = \frac{P}{i\omega\rho} = -\frac{iP}{\omega\rho}$$
(4.50)

which coincides with the result in Eq. 4.48.

Hence, the final result for the velocity of pulsatile flow in a cylindrical tube of radius a is

$$u(r,t) = -\frac{iP}{\omega\rho}\left(1 - \frac{J_0(i^{3/2}\alpha r/a)}{J_0(i^{3/2}\alpha)}\right)e^{i\omega t}.$$
(4.51)

The volumetric flow rate Q, is given by

$$
\begin{aligned}
Q &= \int_0^a u2\pi r\,dr \\
&= -\frac{2\pi iP}{\omega\rho}e^{i\omega t}\left[\int_0^a r\,dr - \frac{1}{J_0(i^{3/2}\alpha)}\int_0^a rJ_0\left(i^{3/2}\alpha r/a\right)\,dr\right] \\
&= -\frac{\pi iP}{\omega\rho}e^{i\omega t}a^2\left[1 - \frac{2}{a^2 J_0(i^{3/2}\alpha)}\int_0^a rJ_0\left(i^{3/2}\alpha r/a\right)\,dr\right].
\end{aligned}
$$
(4.52)

Now consider the term

$$\frac{2}{a^2 J_0(i^{3/2}\alpha)} \int_0^a r J_0\left(i^{3/2}\alpha r/a\right) dr, \tag{4.53}$$

by making the substitutions

$$i^{3/2}\alpha = \beta \quad \text{and} \quad r\beta/a = \theta \tag{4.54}$$

the above term simplifies to

$$\frac{2}{\beta^2 J_0(\beta)} \int_0^\beta \theta J_0(\theta)\, d\theta, \tag{4.55}$$

but

$$\int_0^\beta \theta J_0(\theta)\, d\theta = \beta J_1(\beta). \tag{4.56}$$

Therefore Eq. 4.55 reduces to

$$\frac{2J_1(\beta)}{\beta J_0(\beta)} \tag{4.57}$$

and Eq. 4.52 can be written as

$$Q = -\frac{\pi i P e^{i\omega t}}{\omega \rho} a^2 \chi(\beta), \tag{4.58}$$

where

$$\chi(\beta) = 1 - \frac{2J_1(\beta)}{\beta J_0(\beta)}. \tag{4.59}$$

For small values of β (and hence α) the above expression can be expanded in a Taylor series to give,

$$
\begin{aligned}
\chi(\beta) &= 1 - \frac{2}{\beta} \frac{\left(\frac{\beta}{2} - \left(\frac{\beta}{2}\right)^3/1!2! + \left(\frac{\beta}{2}\right)^5/2!3! - \cdots\right)}{\left(1 - \left(\frac{\beta}{2}\right)^2/(1!)^2 + \left(\frac{\beta}{2}\right)^4/(2!)^2 - \cdots\right)} \\
&= 1 - \frac{1 - \beta^2/8 + \cdots}{1 - \beta^2/4 + \cdots} \\
&= 1 - (1 - \beta^2/8 + \cdots)(1 + \beta^2/4 - \cdots) = -\beta^2/8 + O(\beta^4). \tag{4.60}
\end{aligned}
$$

By substituting $\beta = i^{3/2}\alpha$ into the above equation we get

$$\chi(\beta) = \chi(\alpha) = \frac{i\alpha^2}{8} + O(\alpha^4). \tag{4.61}$$

Then substituting this result into Eq. 4.58 and using the definition of α as given in Eq. 4.44, we finally obtain an expression for Q valid for small α, namely

$$Q = \left(\frac{P\pi a^4}{8\mu} + O(\alpha^4)\right) e^{i\omega t}. \qquad (4.62)$$

Again as $\alpha \to 0$ then $\omega \to 0$ and $Q \to Q_0$ where

$$Q_0 = \frac{\pi P a^4}{8\mu} e^{i\omega t} \qquad (4.63)$$

and

$$|Q_0| = \frac{\pi P a^4}{8\mu} \qquad (4.64)$$

where $|Q_0|$ is the volumetric flow rate for a constant pressure gradient and is the same as for steady Poiseuille's flow.

By separating Eq. 4.58 into its real and imaginary parts we get

$$Q = \frac{\pi P a^2}{\omega\rho} \{[\chi_2(\alpha)\cos\omega t + \chi_1(\alpha)\sin\omega t] - i[\chi_1(\alpha)\cos\omega t - \chi_2(\alpha)\sin\omega t]\} \quad (4.65)$$

where

$$\chi(\alpha) = \chi_1(\alpha) + i\chi_2(\alpha). \qquad (4.66)$$

The real part of Eq. 4.65 gives the flux when the pressure is $P\cos\omega t$ and the imaginary part gives the flux when it is $P\sin\omega t$.

4.6 Further Discussion on Pulsatile Flow

We are interested in two quantitative aspects of pulsatile flow namely,

(i) The magnitude of Q,

$$|Q| = \frac{\pi P a^2}{\omega\rho}\sqrt{\chi_1^2(\alpha) + \chi_2^2(\alpha)} = \frac{\pi P a^4}{\mu\alpha^2}\sqrt{\chi_1^2(\alpha) + \chi_2^2(\alpha)}. \qquad (4.67)$$

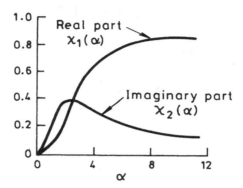

Fig. 4.10: Graphs of $\chi_1(\alpha)$ and $\chi_2(\alpha)$.

(ii) The phase angle between the pressure gradient $Pe^{i\omega t}$ and the flow rate Q, given
by (From Eq. 4.58),

$$\tan\phi = \frac{\chi_1}{\chi_2}. \tag{4.68}$$

From Eq. 4.59,

$$\chi_1 + i\chi_2 = 1 - \frac{2}{i^{3/2}\alpha}\frac{J_1(i^{3/2}\alpha)}{J_0(i^{3/2}\alpha)}. \tag{4.69}$$

The graphs of the values of χ_1 and χ_2 from a table of Bessel functions are given
in Fig. 4.10.

We also have

$$\frac{|Q|}{|Q_0|} = \frac{8\sqrt{\chi_1^2 + \chi_2^2}}{\alpha^2}. \tag{4.70}$$

Since the heart beats about 72 times per minute, we can take

$$\frac{2\pi}{\omega} = \frac{60}{72} \quad \text{or} \quad \omega \approx 8 \,\text{rad sec}^{-1}.$$

For blood, taking $\rho = 1.05$ gm cc^{-1}, $\mu = 4$ cp $= 0.04$ gm cm^{-1}sec^{-1}, and $a = 0.5$ cm,
we get $\alpha \approx 7$. Even for this value, the volumetric flow rate $|Q|$ is about one-eighth of
the steady state value. Relationships of α with the flow rate for various artery sizes
are shown in Table 4.1.

Table 4.1: Relationships of artery size, flow rate and Womersley parameters are shown for the pressure gradient and blood viscosity being taken as 100 gm cm^{-1} sec^{-2} and 0.04 gm cm^{-1}sec^{-1} respectively.

Womersley Parameter, λ	Artery Radius, a (cm)	Flow Rate, $|Q|$ (cm^{3}sec^{-1})
2	0.100	0.083
2	0.200	1.336
2	0.400	21.373
2	0.500	52.180
4	0.100	0.036
4	0.200	0.578
4	0.400	9.251
4	0.500	22.587
6	0.100	0.017
6	0.200	0.276
6	0.400	4.417
6	0.500	10.784
8	0.100	0.010
8	0.200	0.165
8	0.400	2.633
8	0.500	5.429
10	0.100	0.007
10	0.200	0.109
10	0.400	1.737
10	0.500	4.242

Introducing an impedance parameter z defined by

$$z = \frac{-\frac{\partial p}{\partial x}}{Q} = \frac{Pe^{i\omega t}}{Q}, \tag{4.71}$$

which approaches z_0 as $\alpha \rightarrow 0$ (ie., $\omega \rightarrow 0$) where $z_0 = \frac{P}{Q_0}$. We thus have

$$\frac{|z|}{z_0} = \frac{Q_0}{|Q|}. \tag{4.72}$$

The impedance parameter z introduced in Eq. 4.71 is usually called the longitudinal impedance parameter. The idea of impedance is analogous to the concept of transfer function in a mechanical and electrical system. In fact, if the ratio between the pressure gradient and the flow is used to calculate impedance z as defined by Eq. 4.71, it must then be characterized by both a magnitude and an angle, and thus can be written as $z = |z|e^{i\theta}$.

This phase difference, θ, is very important, since it determines the work done by the system, obtained by the product of pressure and rate of flow. The impedance is a function of position, but not a function of time. If it is used to determine the apparent phase velocity then the speed of propagation will also be a function of position. However, it is highly complex to calculate the impedance as it involves a measurement of both pressure and flow. In practice, the pressure gradient-flow relation has usually been taken from the theoretical equations and used in the reverse direction to calculate pulsatile flow.

4.7 The Pulse Wave

It is a well established fact that the flow of blood is pulsatile as a consequence of the beating of the heart. This beating produces a pressure wave that travels through the blood and this pressure wave is the pulse felt in the wrist. In practice, the heart beats about a constant period T of approximately 1 second. We are interested here to obtain the velocity of this pulse wave.

Consider an elastic tube through which the fluid is flowing. Assume now that the fluid is both inviscid and incompressible. Let $A(x,t)$ be the cross-sectional area at a distance x and at any time t, $u(x,t)$ the velocity parallel to the tube, and ρ the density of the fluid. The net force in the positive x direction on the volume element (see Fig. 4.11) is

$$pA - \left[pA + \frac{\partial}{\partial x}(pA)\Delta x\right] + p_0\frac{\partial A}{\partial x}\Delta x = -\frac{\partial}{\partial x}[(p - p_0)A]\Delta x. \qquad (4.73)$$

Now this must be equal to the mass times acceleration of the volume element given by,

$$\rho A\Delta x\left(\frac{\partial u}{\partial t} + u\frac{\partial u}{\partial x}\right). \qquad (4.74)$$

Thus the equation of motion for the fluid through the pipe is

$$\rho A\left(\frac{\partial u}{\partial t} + u\frac{\partial u}{\partial x}\right) = -\frac{\partial}{\partial x}\left((p - p_0)A\right). \qquad (4.75)$$

The continuity equation for an incompressible fluid is

$$\frac{\partial A}{\partial t} + \frac{\partial}{\partial x}(Au) = 0, \qquad (4.76)$$

and the pressure-radius elastic relation is

$$p - p_0 = \frac{Eh}{r_0}\left(1 - \frac{r_0}{r}\right) = \frac{Eh}{r_0}\left(1 - \left(\frac{A_0}{A}\right)^{1/2}\right) \qquad (4.77)$$

where $A = \pi r^2$ and $A_0 = \pi r_0^2$.

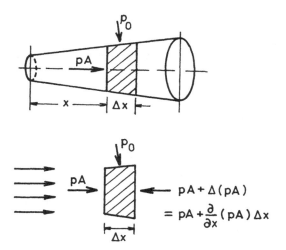

Fig. 4.11: Pressure wave and volume element of fluid flowing in the x-direction.

We linearize Eqs. 4.75-4.77 by assuming that u, $(p - p_0)$ and $(A - A_0)$ (ie., $A = A_0 + h$, where h is some small parameter) together with their derivatives are small. This approach leads to the following three equations:

$$\rho \frac{\partial u}{\partial t} = -\frac{\partial p}{\partial x}, \tag{4.78}$$

$$\frac{\partial A}{\partial t} + A_0 \frac{\partial u}{\partial x} = 0, \tag{4.79}$$

$$p - p_0 = \frac{Eh}{2r_0 A_0}(A - A_0). \tag{4.80}$$

Differentiating Eq. 4.78 with respect to x and Eq. 4.79 with respect to t, we can eliminate u from both equations to give

$$\frac{\partial^2 A}{\partial t^2} = \frac{A_0}{\rho} \frac{\partial^2 p}{\partial x^2}. \tag{4.81}$$

Now from Eq. 4.80 we obtain

$$\frac{\partial^2 p}{\partial t^2} = \frac{Eh}{2r_0 A_0} \frac{\partial^2 A}{\partial t^2}. \tag{4.82}$$

Therefore on combining these two equations we finally obtain

$$\frac{\partial^2 p}{\partial x^2} = \frac{1}{c^2}\frac{\partial^2 p}{\partial t^2} \tag{4.83}$$

where $c^2 = Eh/2\rho r_0$. Equation 4.83 is the classical wave equation with wave speed c. It can be seen that $u(x,t)$ also satisfies the wave equation given by,

$$\frac{\partial^2 u}{\partial x^2} = \frac{1}{c^2}\frac{\partial^2 u}{\partial t^2}. \tag{4.84}$$

We will discuss this equation in the next section.

4.8 Mones-Korteweg Expression for Wave Velocity in an Inviscid Fluid-Filled Elastic Cylindrical Tube

We need to employ the governing fluid-dynamical equations, namely the continuity and momentum equations, the governing equations of tube motion and the matching kinematic boundary conditions. The following assumptions are incorporated in the basic equations:

(i)   Viscosity is neglected.

(ii) The axial flow velocity is small relative to the pulse wave velocity (of the order of less than 10 per cent).

(iii) The vessel diameter is likewise an order of magnitude smaller than the wavelength.

Let u and v denote the flow velocities in the axial and radial directions (x and r) respectively. By examining each equation separately the following results are obtained:

(a) *Continuity Equation*

The continuity equation is given by,

$$\frac{\partial u}{\partial x} + \frac{1}{r}\frac{\partial (rv)}{\partial r} = 0. \tag{4.85}$$

We now introduce the average axial flow velocity

$$\bar{u}(x,t) = \frac{1}{\pi a^2}\int_0^a u(x,r,t)2\pi r\,dr, \tag{4.86}$$

where a is the tube radius.

Upon multiplying both terms of the continuity equation by $2\pi r dr$, integrating with respect to r, and imposing the following boundary conditions:

$$v(x,r,t)|_{r=0} = 0, \quad \text{(due to axisymmetrical flow)} \tag{4.87a}$$

and

$$v(x,r,t)|_{r=a} = v_w, \quad \text{(wall velocity)} \tag{4.87b}$$

we finally obtain,

$$v_w(x,t) = -\frac{a}{2}\frac{\partial \bar{u}(x,t)}{\partial x}. \tag{4.88}$$

(b) *Momentum Equations*

The momentum equations are,

$$\rho\left(\frac{\partial u}{\partial t} + u\frac{\partial u}{\partial x} + v\frac{\partial u}{\partial r}\right) = -\frac{\partial p}{\partial x}, \tag{4.89a}$$

$$\rho\left(\frac{\partial v}{\partial t} + u\frac{\partial v}{\partial x} + v\frac{\partial v}{\partial r}\right) = -\frac{\partial p}{\partial r}. \tag{4.89b}$$

As a result of our assumptions, the radial flow velocity and the convective acceleration terms are respectively of smaller order of magnitude relative to the axial flow velocity and the local acceleration terms. The axial and radial momentum equations now result in:

$$\rho\frac{\partial \bar{u}(x,t)}{\partial t} = -\frac{\partial p(x,t)}{\partial x}, \tag{4.90}$$

$$\frac{\partial p}{\partial r} = 0, \tag{4.91}$$

where ρ is the fluid density and the average axial velocity $\bar{u}$ is as defined in Eq. 4.86.

(c) *Tube Motion Equation*

By considering the dynamic equilibrium of a cylindrical tube element in the radial direction, and by neglecting the inertial force to the motion of the tube wall element (of circumferential and axial dimensions $a d\theta$ and dx respectively, and wall thickness h), we obtain

$$pa d\theta dx - (\sigma h dx)d\theta = 0. \tag{4.92}$$

By putting

$$\sigma = Ee = E(\eta/a), \tag{4.93}$$

where η is the wall displacement, and E the wall material elastic modulus, and simplifying Eqs. 4.92 and 4.93, we get

$$\eta = \frac{a^2}{hE} p(x,t). \tag{4.94}$$

(d) *Kinematic Matching Boundary Conditions*

By requiring that the fluid velocity at the wall, v_w, equals the wall motion velocity ($\dot{\eta}$), and assuming the variation of wall displacement η in the axial direction is negligible we obtain

$$v_w(x,t) = \dot{\eta}(x,t). \tag{4.95}$$

By combining Eqs. 4.88, 4.90, 4.94 and 4.95, we obtain the pulse wave equations, namely,

$$\frac{\partial^2 \bar{u}}{\partial x^2} = \frac{1}{c^2}\frac{\partial^2 \bar{u}}{\partial t^2} \tag{4.96a}$$

and

$$\frac{\partial^2 p}{\partial x^2} = \frac{1}{c^2}\frac{\partial^2 p}{\partial t^2}, \tag{4.96b}$$

where the pulse wave velocity, c, is given by

$$c = \left(\frac{Eh}{2a\rho}\right)^{1/2}. \tag{4.97}$$

4.9 Applications in the Cardiovascular System

The above analysis with the derived expression for the pulse wave velocity can provide the basis for characterizing the modulus of elasticity (Young's modulus) as a function of its inner radius, as well as for demonstrating the construction of the pressure waveform from the flow-velocity waveform and vice versa.

The expression 4.97 for the pulse wave velocity enable the determination of the arterial modulus E (as a function of a) by ultrasound measurement (sector scan echocardiography) of the values of arterial dimensions (a, h), the waveforms of the arterial inner radius at two sites, the transit time (as the time interval between the waveform peaks or centroids), and hence the pulse wave velocity. The present day resolution of sector scan echocardiography systems can realistically enable us to determine only the aortic pressure with reasonable accuracy.

The solution to Eq. 4.96a can be written, in d'Alembert's form as,

$$\bar{u} = f(x - ct) + g(x + ct). \tag{4.98}$$

The functions f and g represent waves travelling in the positive and negative x directions respectively with speed c.

Neglecting the reflected wave, and using Eq. 4.90, we get

$$\bar{u} = f(x - ct) \tag{4.99a}$$

and

$$p = \rho c f(x - ct). \tag{4.99b}$$

Thus, the arterial pressure and velocity are proportional to each other; this explains why the pressure and velocity have similar waveforms. This relation can provide the basis of noninvasive (without any surgical intervention) measurement of arterial pressure waveform from noninvasive pulsed Doppler flowmeter measurement of the averaged blood flow velocity, $\bar{u}$, waveform; and the noninvasive measurement of the pulse wave velocity as mentioned above.

If the $\bar{u}(x,t)$ waveform is measurable by means of the pulsed Doppler flowmeter, then it can be expressed in Fourier series form by computing the Fourier coefficients, u_n, by means of the Euler integral formula; this is illustrated below:

Assume

$$\bar{u}(x,t) = \sum_{n=-\infty}^{\infty} u_n(x)e^{(i2\pi nt)/T} \tag{4.100}$$

such that

$$u_n = \frac{1}{T}\int_0^T \bar{u}(x,t)e^{-(i2\pi nt)/T}\,dt, \tag{4.101}$$

where T is the period of the heart beat or cardiac cycle.

Let the corresponding pressure waveform be given by

$$p(x,t) = \sum_{n=-\infty}^{\infty} p_n(x)e^{(i2\pi nt)/T} \tag{4.102}$$

where

$$p_n = \frac{1}{T}\int_0^T p(x,t)e^{-(i2\pi nt)/T}\,dt, \tag{4.103}$$

Since from Eq. 4.99

$$p_n = \rho c u_n, \tag{4.104}$$

we can construct the pressure waveform $p(x,t)$ corresponding to the pulsed Doppler flowmetry derived $\bar{u}(x,t)$ waveform, by following the steps given below:

(i) Determine p_n, associated with u_n of $\bar{u}(x,t)$ corresponding to each constituent frequency parameter $\omega_n = 2\pi n/T$.

(ii) Obtain the composite waveform $p(x,t)$ according to Eq. 4.102.

(iii) Calibrate the waveform by having the peak waveform amplitude equal to the systolic pressure measured by sphygmomanometry.

Thus, this analysis enable noninvasive and continuous determination of the pressure waveform. By studying the hitherto unavailable aortic pressure waveforms of different diseases of the cardiovascular system, we can associate waveform parameters with cardiovascular abnormalities and develop a new cardiovascular diagnostic field, based on the pressure waveform analysis.

4.10 Wave Propagation Accounting for Viscosity and its Application to Cardiac Output Determination

This theory incorporates the effect of viscosity, μ, in the analysis of pulse wave propagation in viscous fluid-filled cylindrical tubes. The other assumptions stated in Section 4.8 are, however, still relevant.

The equations of continuity and momentum for a viscous fluid are as follows:

$$\frac{\partial u}{\partial x} + \frac{1}{r}\frac{\partial(rv)}{\partial r} = 0, \tag{4.105}$$

$$\rho\frac{\partial u}{\partial t} + \frac{\partial p}{\partial x} = \mu\left(\frac{\partial^2 u}{\partial r^2} + \frac{1}{r}\frac{\partial u}{\partial r}\right), \tag{4.106}$$

$$\frac{\partial p}{\partial r} = 0, \tag{4.107}$$

where u and v are the axial and radial (x and r) flow velocities repectively.

The elastic tube motion equations are (see Fung [20]):

$$\rho_w h\frac{\partial^2 \eta}{\partial t^2} = \sigma_{rr} - \frac{Eh}{1-\nu^2}\left(\frac{\eta}{a^2} + \frac{\nu}{a}\frac{\partial \xi}{\partial x}\right); \quad \sigma_{rr} = p, \tag{4.108}$$

$$\rho_w h\frac{\partial^2 \xi}{\partial t^2} = -\sigma_{rx} - \frac{Eh}{1-\nu^2}\left(\frac{\partial^2 \xi}{\partial x^2} + \frac{\nu}{a}\frac{\partial \eta}{\partial x}\right); \quad \sigma_{rx} = \mu\frac{\partial u}{\partial r}, \tag{4.109}$$

where ρ_w denotes the wall density, E the wall modulus, ν is Poisson's ratio, h is the wall thickness, and (ξ, η) are the axial and radial displacements respectively. Further,

$$\left.\frac{\partial \xi}{\partial t}\right|_{r=a} = u; \quad \left.\frac{\partial \eta}{\partial t}\right|_{r=a} = v. \tag{4.110}$$

The waveforms for axial and radial flow velocities, pressure, and axial and radial wall velocities are represented by:

$$u(x, r, t) = u_0(r)e^{i(kx - \omega t)},$$

$$v(x, r, t) = v_0(r)e^{i(kx - \omega t)},$$

$$p(x, t) = p_0 e^{i(kx - \omega t)},$$

$$\xi(x, t) = \xi_0 e^{i(kx - \omega t)},$$

$$\eta(x, t) = \eta_0 e^{i(kx - \omega t)}, \tag{4.111}$$

where ω $(= 2\pi/T)$ is the frequency, λ $(= 2\pi/k)$ is the wave length and ω/k is the wave speed denoted by c.

The actual waveforms will be a composite of harmonic components with different frequencies. However, for the purpose of analysis, it suffices to concern ourselves with a pulse wave of frequency ω, obtain the relationship between the wave amplitudes of the averaged flow velocity and pressure, and thereby demonstrate how a composite pressure wave can be reconstructed from the composite flow velocity wave, which can be monitored by means of pulsed doppler flowmetry.

Substituting the expressions for u and p in Eq. 4.106, we get the following non-homogeneous Bessel equation in u_0:

$$\frac{d^2 u_0}{dr^2} + \frac{1}{r}\frac{du_0}{dr} + \frac{i\omega\rho}{\mu}u_0 = \frac{ikp_0}{\mu}, \tag{4.112}$$

whose solution is given by

$$u_0(r) = \frac{k}{\omega\rho}p_0 + AJ_0(\beta r) + BY_0(\beta r); \quad \beta = \frac{i\omega\rho}{\mu}, \tag{4.113}$$

in which the term $Y_0(\beta r)$ becomes irregular at $r = 0$, and hence, must be omitted.

Likewise, we substitute the waveform expressions for u and v in Eq. 4.105, integrate the resulting differential equation:

$$\frac{d(rv_0)}{dr} = -ik(ru_0), \tag{4.114}$$

substitute for $u_0(r)$ from Eq. 4.113, and by keeping in mind that $v_0|_{r=0} = 0$, (by virtue of axial symmetry) and that $\int rJ_0(r)dr = rJ_1$, we obtain:

$$v_0(r) = -\frac{ik^2}{2\omega\rho}p_0 r - \frac{ikA}{\beta}J_1(\beta r). \tag{4.115}$$

Next, we substitute the waveform expressions 4.111 in the dynamic equations 4.108 and 4.109 of the tube wall segment, and obtain,

$$p_0 = \frac{Eh}{(1 - \nu^2)a^2}\eta_0 + \frac{iEhk\nu}{(1 - \nu^2)a}\xi_0 - \rho_w h\omega^2\eta_0, \tag{4.116}$$

$$\xi_0 = \frac{i\nu}{ak}\eta_0 + \frac{(1-\nu^2)\mu A\beta}{Ehk^2}J_1(\beta a) - \frac{\rho_w h\omega^2(1-\nu^2)}{Ehk^2}\xi_0. \tag{4.117}$$

The last terms in Eqs. 4.116 and 4.117 are of relatively small order of magnitude, and may thus be neglected.

Next by invoking the boundary conditions 4.110:

$$-i\omega\xi_0 = u_0(a); \qquad -i\omega\eta_0 = v_0(a), \tag{4.118}$$

wherein $u_0(a)$ and $v_0(a)$ are given by 4.113 and 4.115, so that

$$\xi_0 = \frac{-k}{i\omega^2\rho}p_0 - \frac{A}{i\omega}J_0(\beta a), \tag{4.119}$$

$$\eta_0 = \frac{k^2 a p_0}{2\omega^2\rho} + \frac{Ak}{\beta\omega}J_1(\beta a), \tag{4.120}$$

and by substituting the above equations into Eqs. 4.116 and 4.117, we obtain:

$$p_0\left(\frac{E^*}{\rho}\frac{k^2}{\omega^2}\left(\frac{1}{2}-\nu\right)-1\right) + AE^*\frac{k}{\omega}\left(\frac{J_1(\beta a)}{a\beta}-\nu J_0(\beta a)\right) = 0, \tag{4.121}$$

and

$$p_0\left(\frac{k}{2\omega(i\omega)\rho}(2-\nu)\right) + \frac{A}{\omega^2}\left(\left(\frac{\mu\beta\omega^2}{E^*ak^2}+\frac{i\omega\nu}{a\beta}\right)J_1(\beta a)+\omega\frac{J_0(\beta a)}{i}\right) = 0, \tag{4.122}$$

wherein $E^* = Eh/(1-\nu^2)a$.

For a non-zero wave amplitude, the determinant of the above set of linear equations in p_0 and A must be zero, which finally yields the following characteristic equation in $\underline{X} = \frac{k^2}{\omega^2}E^*$, after some mathematical simplification:

$$\left((1-2\nu)\underline{X}-2\rho\right)\left(\left(\frac{\nu}{a\beta}-S\right)\underline{X}+\frac{\rho}{a}\right) + \frac{(2-\nu)}{2\rho}\left(\frac{1}{a\beta}-S\nu\right)\underline{X}^2 = 0, \tag{4.123}$$

where

$$S = \frac{J_0(\beta a)}{J_1(\beta a)}. \tag{4.124}$$

From this equation we can determine the value of the parameter $\underline{X}$ in terms of the known values a, ν, S and β. The above analysis will now be applied to determine the cardiac output (CO).

Now ω/k, the wave speed, can be evaluated by monitoring the transit time as the time interval between the peaks or centroids of ultrasonically measured waveforms of the arterial diameter at two arterial sites at a known distance apart. Then, since the value of $\underline{X}$ can be determined from Eq. 4.123, that of E^* can also be calculated.

Having evaluated E^* and ω/k, we can now compute the value of A in terms of p_0 from either Eq. 4.121 or 4.122 as,

$$A = \frac{-\left(\frac{E^* k^2}{2\rho\omega^2}(1-2\nu)-1\right)p_0}{\frac{E^* k}{\omega}\left(\frac{J_1(\beta a)}{a\beta} - \nu J_0(\beta a)\right)}, \tag{4.125}$$

and thereby express $u_0(r)$ in Eq. 4.113 completely in terms of p_0 as follows:

$$u_0(r) = \left(\frac{k}{\omega\rho} - \theta J_0(\beta r)\right)p_0, \tag{4.126}$$

where

$$\theta = \frac{\left(\frac{E^* k^2(1-2\nu)}{2\rho\omega^2}-1\right)}{E^* \frac{k}{\omega}\left(\frac{J_1(\beta a)}{a\beta} - \nu J_0(\beta a)\right)}. \tag{4.127}$$

The above equation enables the lumen averaged velocity amplitude, $\bar{u}_0$ (corresponding to the pulse frequency ω), given by

$$\bar{u}_0 = \frac{1}{\pi a^2}\int_0^a u_0(r)2\pi r\, dr, \tag{4.128}$$

to be related to the corresponding pressure amplitude, p_0, in terms of quantities that are known (such as ρ), monitorable (such as a), and computable (such as S and β). We can now adopt the following steps to construct the arterial pressure waveform:

(i) By means of pulsed Doppler flowmetry, monitor (at an arterial site $x = x_0$), the lumen averaged flow velocity waveform, namely

$$\bar{u}(x_0, t) = \sum_{n=-\infty}^{\infty} \bar{u}_n e^{in\omega t}. \tag{4.129}$$

(ii) Determine the spectral components, given by,

$$\bar{u}_n = \frac{1}{T}\int_0^T \bar{u}(t)e^{-in\omega t}\, dt. \tag{4.130}$$

(iii) For each harmonic component $\bar{u}_n$ with frequency $\omega_n = n\omega$, we can determine the associated p_n from Eqs. 4.126 and 4.128, as

$$\begin{aligned}
p_n &= \frac{\pi a^2 \bar{u}_n}{\int_0^a \left(\frac{k}{\omega_n\rho} - \theta J_0(\beta r)\right)2\pi r\, dr} \\
&= \frac{a\bar{u}_n}{\left(\frac{ak}{\omega_n\rho} - \frac{2\theta}{\beta}J_1(\beta a)\right)}
\end{aligned} \tag{4.131}$$

(iv) Obtain the pressure waveform (at a certain $x = x_0$), which is given by,

$$p(x_0, t) = \sum_{n=-\infty}^{\infty} p_n e^{i\omega_n t}. \tag{4.132}$$

(v) Finally calibrate $p(x_0, t)$ by making the peak cyclic value equals to the systolic arterial pressure value obtained by means of sphygmomanometry, which amounts to shifting the $p(x_0, t)$ waveform upward by its mean value amount.

We can also express the ascending aortic flow rate, Q, and hence, the cardic ouput, CO, in terms of p_0, k and ω, ie., in terms of the parameters of the ascending aortic pressure waveform.

The cardiac output is given by,

$$CO = \int_0^T Q \, dt, \tag{4.133}$$

where

$$Q(t) = \pi a^2 \bar{u}(x_0, t);$$

$$\bar{u}(x_0, t) = \sum_{n=-\infty}^{\infty} \bar{u}_n e^{in\omega t};$$

$$\bar{u}_n = \frac{1}{\pi a^2} \int_0^a u_n(r) 2\pi r \, dr;$$

$$u_n(r) = \left(\frac{k}{\omega_n \rho} - \theta J_0(\beta r) \right) p_n; \quad \omega_n = n\omega,$$

$$p_n = \frac{1}{T} \int_0^T p(x_0, t) e^{-i\omega_n t} \, dt. \tag{4.134}$$

Hence, the cardiac output can be calculated knowing the values of the measurable parameters.

4.11 Flow Through a Converging-Diverging Duct

We will now study the flow of an incompressible fluid through an axisymmetric converging-diverging tube. The analysis will be carried out in the context of a mathematical model for mild *stenosis*.

In the arterial systems of humans or animals, it is quite common to find localized narrowings, commonly called stenoses, caused by intravascular plaques. These stenoses disturb the normal pattern of blood flow through the artery. A knowledge of the flow characteristics in the vicinity of a stenosis may help to further understand some major complications which can arise such as, an ingrowth of tissue in the artery, the development of a coronary thrombosis, the weakening and bulging of the artery downstream from the stenosis, etc.

Investigation of the role of hydrodynamic factors in the development and progression of the above complications provides relevance to this analysis of flow through a model arterial stenosis. Conversely, the expression and resulting nomograms for the increased mean pressure drop through a stenotic segment (averaged over the period of the heart cycle, T), in terms of the vessel size and cardiac output, can provide us an index of the degree of stenosis.

It has been observed that at certain arterial sites, stagnation zones occur, which are due to the separation of the main flow from the walls of the arteries. It has been suggested that stagnation zones near prosthetic heart valves contribute to the formation of a thrombus. Thus, it seems reasonable to speculate that if separation regions of relatively stagnant flow occur in stenotic segments, they may well contribute to thrombosis.

The relationship between the mean pressure, p_0, averaged over the cardiac cycle period, T, (whereby $p_0 = (1/T) \int_0^T p\, dt$) and mean flow, q_0 ($= (1/T) \int_0^T q\, dt$), can be modeled as the Poiseuille (steady) flow of a viscous and incompressible Newtonian fluid (of viscosity μ, taken as 0.04 Poise, or dyne sec cm^{-1}, for blood at 37°C) through a straight circular rigid tube of radius a. The familiar relationship between q_0 and the gradient of p_0 is given by,

$$q_0 = -\frac{\pi a^4}{8\mu} \frac{dp_0}{dx}. \tag{4.135}$$

The associated axial velocity profile is parabolic and given by,

$$u(r) = -\frac{1}{4\mu} \frac{dp_0}{dx}(a^2 - r^2). \tag{4.136}$$

However, if a converging-diverging segment is placed in a straight tube, the velocity profile will be altered, with a subsequent change in other flow characteristics. In order to assess the mean pressure drop due to a stenosis of measurable geometry under a pulsed Doppler based monitorable mean (lumen averaged) flow velocity, the specific problem considered here will be that of steady axisymmetric flow of an incompressible fluid through a converging-diverging tube with circular cross-section having rigid walls (Young [74]; Forrester and Young [16]).

Figure 4.12 illustrates the geometry of a collar-like stenosis located in a relatively straight artery. Consider that the axial coordinate and velocity are to be denoted as $\hat{z}$ and $\hat{u}$ respectively and the radial coordinate and velocity are to be similarly denoted as $\hat{r}$ and $\hat{v}$ respectively. The local radius of the axisymmetric tube is $\hat{R}$, whereas R_0 is the radius of the artery proximal to the stenoised segment.

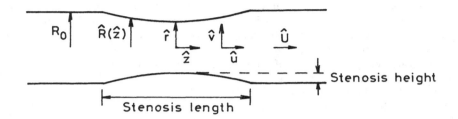

Fig. 4.12: Collar-like stenosis in a relatively straight artery.

The governing equations are the continuity equation and the Navier-Stokes equations, wherein we will first introduce the following dimensionless variables:

$$\left.\begin{array}{l} r = \dfrac{\hat{r}}{R_0}, \quad z = \dfrac{\hat{z}}{R_0}, \quad R = \dfrac{\hat{R}}{R_0}, \quad u = \dfrac{\hat{u}}{\overline{U}_0}, \\[3mm] v = \dfrac{\hat{v}}{\overline{U}_0}, \quad p = \dfrac{\hat{p}}{\rho \overline{U}_0^2}, \quad U = \dfrac{\hat{U}}{\overline{U}_0}, \end{array}\right\} \tag{4.137}$$

where $\hat{U}$ is the centreline velocity, $\overline{U}_0$ is the lumen averaged velocity in the unobstructed tube, $\hat{p}$ is the pressure and ρ is the fluid density. It is to be noted when applying the analysis to an arterial segment, that the values of R_0 and $\overline{U}_0$ will be time averaged over a cardiac cycle period.

The continuity equation and the Navier-Stokes equations (in the axial and radial directions) can be written in dimensionless form as follows,

$$\frac{\partial u}{\partial z} + \frac{\partial v}{\partial r} + \frac{v}{r} = 0, \tag{4.138}$$

$$u\frac{\partial u}{\partial z} + v\frac{\partial u}{\partial r} + \frac{\partial p}{\partial z} = \frac{2}{Re}\left(\frac{\partial^2 u}{\partial r^2} + \frac{1}{r}\frac{\partial u}{\partial r} + \frac{\partial^2 u}{\partial z^2}\right), \tag{4.139}$$

$$u\frac{\partial v}{\partial z} + v\frac{\partial v}{\partial r} + \frac{\partial p}{\partial r} = \frac{2}{R_e}\left(\frac{\partial^2 v}{\partial r^2} + \frac{1}{r}\frac{\partial v}{\partial r} - \frac{v}{r^2} + \frac{\partial^2 v}{\partial z^2}\right), \tag{4.140}$$

where R_e denotes the Reynolds number $(2R_0\overline{U}_0\rho/\mu)$ and μ denotes the fluid viscosity. These equations are non-linear and exact solutions are not easy to obtain.

However, approximate solutions can be obtained on the basis of some simplifying assumptions (Morgan and Young [46]).

Integrating Eq. 4.139 over the cross-section of the tube, and imposing the non-slip boundary condition,

$$u = v = 0$$

at the wall, the integral momentum equation is obtained as,

$$\frac{1}{2}\frac{\partial}{\partial z}\int_0^R ru^2\,dr = -\int_0^R r\frac{\partial p}{\partial z}\,dr + \frac{2}{R_e}\left(R\left.\frac{\partial u}{\partial r}\right|_R + \int_0^R r\frac{\partial^2 u}{\partial z^2}\,dr\right). \tag{4.141}$$

Likewise, by first multiplying Eq. 4.139 by ru, and then integrating from $r = 0$ to $r = R$, we obtain,

$$\frac{1}{3}\frac{\partial}{\partial z}\int_0^R ru^3\,dr = -\int_0^R ru\frac{\partial p}{\partial z}\,dr + \frac{2}{R_e}\left(-\int_0^R r\left(\frac{\partial u}{\partial r}\right)^2\,dr + \int_0^R ru\frac{\partial^2 u}{\partial z^2}\,dr\right). \tag{4.142}$$

In Eqs. 4.141 and 4.142, the terms containing $\partial^2 u/\partial z^2$ associated with the viscous component of the normal stress in the axial direction are considered negligible. Further, the $\int_0^R r\frac{\partial p}{\partial z}\,dr$ term appearing in Eq. 4.141 can be reasonably approximated as

$$\int_0^R r\frac{\partial p}{\partial z}\,dr \approx R^2\int_0^R ru\frac{\partial p}{\partial z}\,dr, \tag{4.143}$$

assuming in a constriction the pressure gradient is nearly constant and the velocity profile tends to become flattened.

Invoking these assumptions, Eqs. 4.141 and 4.142 are combined into a single equation in terms of the non-dimensional axial velocity u:

$$\frac{1}{3}R^2\frac{\partial}{\partial z}\int_0^R ru^3\,dr - \frac{1}{2}\frac{\partial}{\partial z}\int_0^R ru^2\,dr = -\frac{2}{R_e}\left(R^2\int_0^R r\left(\frac{\partial u}{\partial r}\right)^2\,dr + R\left.\frac{\partial u}{\partial r}\right|_R\right). \tag{4.144}$$

At this stage, we need to assume the axial velocity profile function, in order to fulfill the following characteristics of flow through the converging-diverging cross-section:

(i) At high Reynolds number, the profile must yield a thin region of high shear near the wall in the converging section, with a flat profile in the core.

(ii) The profile must assume the characteristics of a central jet with low shear, with backflow near the wall, in the diverging section.

(iii) The no slip condition,

$$u = 0 \quad \text{at} \quad r = R.$$

(iv) At the centreline,

$$u = U \quad \text{at} \quad r = 0.$$

(v) The smoothness condition associated with zero stress jump at the origin,

$$\frac{\partial u}{\partial r} = 0 \quad \text{at} \quad r = 0.$$

(vi) By eliminating the pressure term between Eqs. 4.139 and 4.140, and considering the resulting equation in the limit as r approaches zero, yields

$$\frac{\partial^3 u}{\partial r^3} = 0 \quad \text{at} \quad r = 0.$$

(vii) In order that the net flow through any cross-section be a constant,

$$\int_0^R r u \, dr = \frac{1}{2}.$$

The velocity profile that satisfies the above requirements is given in two parts:

(a)

$$u = U \left(A + B \left(\frac{r}{R} \right) + C \left(\frac{r}{R} \right)^2 + D \left(\frac{r}{R} \right)^3 + E \left(\frac{r}{R} \right)^4 \right) \tag{4.145}$$

and the unknown coefficients are obtained by satisfying the above constraints (iii) - (vii) to give,

$$u = \frac{1}{R^2} \left\{ R^2 U + 2(3 - 2R^2 U) \left(\frac{r}{R} \right)^2 - 3(2 - R^2 U) \left(\frac{r}{R} \right)^4 \right\}, \text{ for } R^2 U \geq 1.5. \tag{4.146}$$

Thus, the velocity profile is determined in terms of centreline velocity U. It is interesting to note that when $R^2 U = 2$, Eq. 4.146 reduces to $u = \frac{2}{R^2}[1 - (r/R)^2]$, which is a parabolic profile corresponding to the Poiseuille flow.

(b)

$$u = \begin{cases} U, & \text{for } 0 \leq \left(\frac{r}{R} \right) \leq \lambda; \\ \\ a + b \left(\frac{r}{R} \right)^2 + c \left(\frac{r}{R} \right)^4, & \text{for } \lambda < \frac{r}{R} \leq 1, \end{cases} \tag{4.147}$$

which by satisfying the no-slip condition, and $u = U$ and $\frac{\partial u}{\partial r}$ at $r/R = \lambda$, yields

$$u = \begin{cases} U, & \text{for } 0 \leq \left(\frac{r}{R} \right) \leq \lambda; \\ \\ \frac{U}{(1 - \lambda^2)^2} \left(1 - 2\lambda^2 + 2\lambda^2 \left(\frac{r}{R} \right)^2 - \left(\frac{r}{R} \right)^4 \right), & \text{for } \lambda < \left(\frac{r}{R} \right) \leq 1, \end{cases} \tag{4.148}$$

with

$$\lambda = \left(\frac{3}{R^2 U} - 2\right)^{1/2}, \quad \text{and for } R^2 U \le 1.5.$$

The resulting profiles for various values of $R^2 U$ are shown in Fig. 4.13.

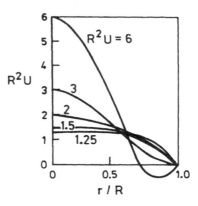

Fig. 4.13: Velocity profile shapes.

Upon substituting the profiles 4.146 and 4.148 into Eq. 4.144, we obtain

$$\frac{dU}{dz} = \left\{-\frac{1}{R^3}\frac{dR}{dz}\left(0.0286(R^2 U)^3 - 0.133(R^2 U)^2 + 0.257 R^2 U - 0.34\right)\right.$$

$$\left. + \frac{4U\Gamma}{Re}\right\}\left(0.0429(R^2 U)^2 - 0.004(R^2 U) - 0.029\right)^{-1}, \qquad (4.149)$$

where

$$\Gamma = \begin{cases} 2 - R^2 U, & \text{for } R^2 U \ge 1.5; \\[2mm] \dfrac{3R^2 U - 1}{9(R^2 U - 1)}, & \text{for } R^2 U < 1.5, \end{cases} \qquad (4.150)$$

The initial condition required to numerically solve this nonlinear differential equation is,

$$\text{at} \quad z = -\infty, \quad R = 1 \quad \text{and} \quad U = 2.$$

What is of interest to us in being able to lead to a noninvasive diagnostic method for detecting the location and degree of stenosis, is the expression for the average pressure gradient given by

$$\left.\frac{\partial p}{\partial z}\right|_{av} = \frac{\int_0^R r\frac{\partial p}{\partial z}\,dr}{\int_0^R r\,dr} = \frac{2}{R^2}\int_0^R r\frac{\partial p}{\partial z}\,dr \qquad (4.151)$$

wherein, from Eq. 4.141

(i)

$$\int_0^R r \frac{\partial p}{\partial z} \, dr = \frac{2}{Re} \left(R \frac{\partial u}{\partial r} \bigg|_R + \int_0^R r \frac{\partial^2 u}{\partial z^2} \, dr \right) - \frac{1}{2} \frac{\partial}{\partial z} \int_0^R r u^2 \, dr \qquad (4.152)$$

(ii) u is given, by Eq. 4.146 and 4.148, in terms of U.

(iii) U is solved by means of Eq. 4.149, for stenosis geometry such as that given by

$$R(z) = \begin{cases} 1 - \frac{\delta}{2} \left(1 + \cos \left(\frac{\pi z}{Z_0} \right) \right), & \text{for} |z| \le Z_0; \\ 1, & \text{for} |z| > Z_0, \end{cases} \qquad (4.153)$$

where δ is the (stenosis or plaque height)$/R_0$ and Z_0 is the (stenosis length)$/2R_0$.

By means of Eqs. 4.146-4.150 and 4.153 the non-dimensional pressure drop Δp can be numerically computed and plotted as a function of δ, Z_0 and R_0. For this purpose using a set of values of (δ, Z_0, R_0) we numerically formulate the shape function $R(z)$ as a function of the non-dimensional coordinate z, solve Eq. 4.149 for U as a function of z, substitute the values of $U(z)$ in Eqs. 4.146 and 4.148 and formulate u as a function of r and z, then employ Eqs. 4.151 and 4.152 to compute the values of $\partial p/\partial z$ for different values of z, and finally determine the value of the pressure drop,

$$\Delta p = \int_{-Z_0}^{Z_0} \frac{\partial p}{\partial z} \, dz. \qquad (4.154)$$

Thus, from the computed nomograms (Fig. 4.14), the non-dimensional pressure drop Δp can be determined for known values of Z_0 (the non-dimensional distance between arterial measurement sites proximal and distal to the stenotic site), δ (the non-dimensional parameter for the degree of stenosis) and R_0 (the radius of the unconstrained sections upstream and downstream of the stenotic site).

Consider now a possible methodology for practical implementation of this analysis, to noninvasively determine the pressure drop Δp across the stenotic segment. By means of a pulse Doppler flowmeter, we can determine the lumen averaged instantaneous flow velocity proximal to the stenotic segment in terms of the mean Doppler frequency shift (Baker and Daigle [4]), as

$$\overline{U} = \frac{c \Delta f}{2 f_0 \cos \theta}; \qquad (4.155)$$

wherein Δf is the mean Doppler frequency shift (averaged over the vessel lumen), c the velocity of sound, f_0 the centre frequency and θ the angle between the velocity and beam vectors. Therefore we can determine the averaged flow velocity $\overline{U_0}$ over a period T, as

$$\overline{U}_0 = \frac{1}{T} \int_0^T \overline{U}(t) \, dt. \qquad (4.156)$$

We then obtain real-time longitudinal cross-sectional views of the stenotic segment by means of a pulse echo sector-scan unit, measure from there the value of the stenosis length, Z_0, as well as the instantaneous values of the stenosis height and the unconstrained tube radius, average them over the cardiac period to obtain the averaged values of R_0 and the stenosis height parameter δ, as defined by Eq. 4.153. Having measured the values of δ, R_0 and Z_0, we can then obtain the value of Δp from the developed nomograms of "Δp vs δ and Z_0", of Fig. 4.14.

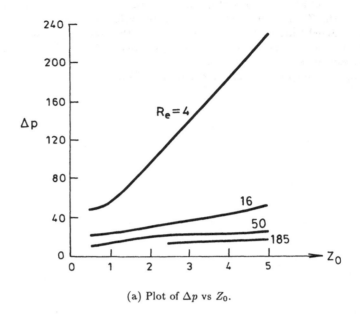

(a) Plot of Δp vs Z_0.

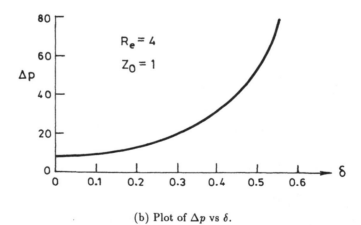

(b) Plot of Δp vs δ.

Fig. 4.14: Plots of Δp.

Chapter 5

NON-NEWTONIAN FLUIDS

5.1 General Introduction

The study of fluid dynamics began with imaginary 'ideal' or 'perfect' fluid that is incompressible and without viscosity or elasticity. Shearing stresses are absent during shear as the fluid movement is completely frictionless. Detailed mathematical relationships have been obtained for the behaviour of a perfect fluid in a wide variety of physical situations. Some of these results have proved to be useful approximations to the performance of real fluids in certain special cases.

However, severe limitations in the practical application of frictionless flow theory to real situations in general led to the development of a dynamical theory for the simplest class of real fluids — Newtonian fluids.

A Newtonian fluid, by definition, is one in which the coefficient of viscosity is constant at all rates of shear. Homogeneous liquids may behave closely like Newtonian fluids. However, there are fluids that do not obey the linear relationship between shear stress and shear strain rate. Fluids that exhibit a non-linear relationship between the shear stress and the rate of shear strain are called *non-Newtonian* fluids. Thick liquids are generally in this category. Many common fluids are non-Newtonian, eg. paints, enamels, varnish, wet clay and mud, solutions of various polymers, suspensions of particles, emulsions of oil in water, etc.

Although the properties of non-Newtonian fluids do not lend themselves to precise and simple analysis that has been developed for Newtonian fluids, the flow of non-Newtonian fluids does possess some interesting and useful characteristics.

We have seen that blood deviates in behaviour from Newtonian fluids. Such anomalous behaviour of blood is often referred to as non-Newtonian properties of blood. These properties are of two types as follows:

(i) at low shear rates, the apparent viscosity increases markedly — sometimes even a certain "yield stress" is required for flow.

(ii) in small tubes, the apparent viscosity at higher rates of shear is smaller than it is in larger tubes.

These two types of anomalies are often referred to as "low shear" and "high shear" effects respectively. It is thus concluded that the behaviour of blood is almost Newtonian at high shear rate, while at low shear rate the blood exhibits yield stress and non-Newtonian behaviour. Most body fluids are non-Newtonian.

5.2 Classification of Non-Newtonian Fluids

Fluids which do not obey the linear relationship between shear stress and rate of shear strain can be grouped into three general classifications:

(1) The simplest type of non-Newtonian fluids is the time independent non-Newtonian fluid in which the shear strain rate is a non-linear function of the shear stress, independent of shearing time and previous shear stress rate history of that fluid.

(2) Time dependent non-Newtonian fluids have more complex shearing stress strain rate relationships. The shearing strain rates in these fluids depend on shearing times or on the previous shear stress rate histories of the fluids and, thus, are not single valued functions of the shear stresses.

(3) In *viscoelastic fluids*, some of the energy of deformation may be recoverable as it is in elastic solids. This is different from truly viscous fluids in which all the energy of deformation is dissipated. Hence, the shear strain as well as shear strain rate are related in some way to shear stress.

5.3 Time Independent Fluids

For a time independent non-Newtonian fluid, the constitutive equation is

$$\tau = f(\dot{\gamma})$$

or,

$$\dot{\gamma} = f(\tau). \tag{5.1}$$

A Newtonian fluid is just a special case of non-Newtonian fluid where the function $f(\dot{\gamma})$ is linear. Figure 5.1 shows some typical shear stress strain rate relations for non-Newtonian fluids.

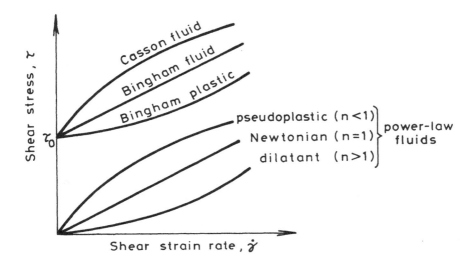

Fig. 5.1: Typical shear stress strain rate relationships for non-Newtonian fluids.

5.3.1 *Power-Law Fluids*

One important class of non-Newtonian fluids is that of power-law fluids which have constitutive equation

$$\tau = \mu\dot{\gamma}^n$$

$$= \mu\dot{\gamma}^{n-1}\dot{\gamma}. \qquad (5.2)$$

This class of non-Newtonian fluids has effective viscosity coefficient or apparent viscosity $\mu\dot{\gamma}^{n-1}$ and does not have a yield stress. If $n < 1$, we get a pseudoplastic power-law fluid characterized by a progressively decreasing apparent viscosity with strain rate. If $n > 1$, we have a dilatant power-law fluid in which the apparent viscosity increases progressively with increasing strain rate. If $n = 1$, we obtain the Newtonian fluid as a special case. At very high shear rates in a real fluid, the apparent viscosity becomes constant and equal to μ_∞, and the shear stress shear rate relationship becomes linear.

Several objections are raised against the power-law model, one of them being for a pseudoplastic fluid, the apparent viscosity is infinite when the shear strain rate is zero. However, this model has been found desirable because of its simplicity, as well as being adequate to analyse such flows as *Couette flow* and flow in pipes and channels. We will discuss this later.

5.3.2 *Bingham Fluids*

Another important non-Newtonian fluid, the Bingham fluid, has constitutive equation

$$\tau = \mu\dot{\gamma} + \tau_0, \quad \tau \geq \tau_0$$
$$\dot{\gamma} = 0. \quad \tau \leq \tau_0 \qquad (5.3)$$

It is also known as Bingham plastic.

It exhibits a yield stress τ_0 at zero shear rate, followed by a linear relationship between shear stress and shear strain rate, the plastic viscosity μ being the slope of the straight line. If $\tau < \tau_0$, no flow takes place. The behaviour of many real fluids such as slurries, household paint and plastics very closely approximate this concept.

5.3.3 *Other Special Non-Newtonian Fluids*

The following are some other empirical equations for special non-Newtonian fluids:

Herschel-Bulkley fluid:

$$\tau = \mu\dot{\gamma}^n + \tau_0, \quad \tau \geq \tau_0$$
$$\dot{\gamma} = 0, \quad \tau \leq \tau_0 \qquad (5.4)$$

Casson fluid:

$$\tau^{\frac{1}{2}} = \mu^{\frac{1}{2}}\dot{\gamma}^{\frac{1}{2}} + \tau_0^{\frac{1}{2}}, \quad \tau \geq \tau_0$$
$$\dot{\gamma} = 0, \quad \tau \leq \tau_0 \qquad (5.5)$$

Prandtl fluid:

$$\tau = A \arcsin\left(\frac{\dot{\gamma}}{C}\right) \qquad (5.6)$$

Prandtl-Eyring fluid:

$$\tau = A\dot{\gamma} + B \arcsin\left(\frac{\dot{\gamma}}{C}\right) \qquad (5.7)$$

Power-Eyring fluid:

$$\tau = A\dot{\gamma} + B \sinh^{-1}(C\dot{\gamma}) \qquad (5.8)$$

Ellis fluid:

$$\dot{\gamma} = A\tau + B\tau^n \qquad (5.9)$$

Reiner-Philliphoff fluid:

$$\tau = \left(\frac{\mu_0 - \mu_\infty}{1 + \left(\frac{\tau}{\tau_0}\right)^2} \right) \dot{\gamma} + \mu_\infty \tag{5.10}$$

Rabinowitsch fluid:

$$\dot{\gamma} = \frac{1}{\mu_0}\tau + \sum_q s_q \tau^{2q+1}. \tag{5.11}$$

The above list of non-Newtonian fluids is by no means complete. However, for biofluid flows, Eqs. 5.2, 5.4 and 5.5 are especially useful.

5.4 Time Dependent Fluids

Some fluids are more complex than those just described and the apparent viscosity depends not only on the strain rate but also on the time the shear has been applied. These can generally be classified into two classes:

(i) thixotropic fluids

(ii) rheopectic fluids.

For a thixotropic fluid, the shear stress decreases with time as the fluid is sheared while for a rheopectic fluid, the shear stress increases with time as the fluid is sheared. An example of a thixotropic fluid is printer's ink.

5.5 Viscoelastic Fluids

A viscoelastic material exhibits both elastic and viscous properties. The simplest viscoelastic fluid is one which is Newtonian in viscosity and obeys Hooke's law for the elastic part, giving the constitutive equation

$$\dot{\gamma} = \frac{\tau}{\mu_0} + \frac{\dot{\tau}}{\lambda} \tag{5.12}$$

where λ is a rigidity modulus.

Rather complex models of viscoelastic materials have been developed in which higher time derivatives of τ and γ appear. For time varying processes, the elastic constants may be complex functions of frequency.

5.6 Laminar Flow of Non-Newtonian Fluids

Let us now discuss a number of non-Newtonian fluid models for laminar flow in a tube and comment on how well they compare to experimental results. The assumptions of no external forces and steady, uniaxial, axisymmetric flow in a rigid, cylindrical tube still apply. It is further assumed that the flow is fully-developed far away from the ends of the tube.

5.6.1 *Power-Law Model*

A simple model taking into account the non-Newtonian behaviour of fluid flow is the power-law model. It suggests a constitutive equation of the form

$$\tau = \mu \dot{\gamma}^n \tag{5.13}$$

where τ is the shear stress

$\dot{\gamma}$ is the shear strain rate.

At this point, since we are dealing with variable viscosity, we shall define what we mean by "variable". If the viscosity is constant, as in the Newtonian case, then it is simply the ratio of stress (τ) to shear rate ($\dot{\gamma}$). However, for variable viscosity, we have "differential" viscosity μ' being defined by the slope of the curve, ie.

$$\mu' = \tan^{-1}\left(\frac{d\tau}{d\dot{\gamma}}\right).$$

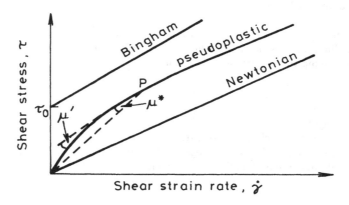

Fig. 5.2: Differential and generalized viscosity for power-law model.

Alternatively, we may consider "generalized" viscosity μ^*, which is the ratio of applied stress to shear rate (see Fig. 5.2). Thus, at point P,

$$\mu^* = \tan^{-1}\left(\frac{\tau}{\dot{\gamma}}\right).$$

This is also called "secant viscosity".

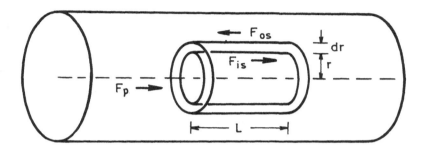

Fig. 5.3: The control volume and the forces acting on its surfaces.

Let us first determine the flux of the flow. Consider a control volume bounded by two coaxial cylinders of radii r and $r + dr$ and both of length L (see Fig. 5.3).

Basically, two types of forces act on this control volume; forces due to the pressure gradient and forces due to the shear stresses.

The force due to a constant pressure gradient of $-P$ across a cylindrical tube of length L is

$$F_p = P \times 2\pi r L \, dr. \tag{5.14}$$

An object under shear stress has equal and opposite forces applied across its opposite faces. Thus there are shear stresses acting both on the inner and outer cylindrical surfaces of the control volume. Let the stress at a distance r from the axis be $\tau(r)$. Then the force acting on the inner cylindrical surface of the volume will be

$$F_{is} = 2\pi r L \tau(r) \tag{5.15}$$

and that on the outer cylindrical surface is

$$
\begin{aligned}
F_{os} &= 2\pi(r+dr)L\tau(r+dr) \\
&= 2\pi(r+dr)L\left[\tau + dr\frac{d\tau}{dr} + O\left((dr)^2\right)\right] \\
&= 2\pi L\left[r\tau + \tau dr + r\,dr\frac{d\tau}{dr} + O\left((dr)^2\right)\right]
\end{aligned}
$$

$$\simeq \ 2\pi L \left[r\tau + \frac{d}{dr}(r\tau)\, dr \right]. \tag{5.16}$$

Balancing forces parallel to the axis,

$$F_{os} - F_{is} \ = \ F_p$$

$$2\pi L\, dr \frac{d}{dr}(r\tau) \ = \ P \times 2\pi r L\, dr$$

$$\frac{d}{dr}(r\tau) \ = \ Pr. \tag{5.17}$$

Integrating Eq. 5.17, we have

$$\tau \ = \ \frac{Pr}{2} + \frac{A}{r}. \tag{5.18}$$

For stress to be finite on the axis ($r = 0$), we need $A = 0$. Therefore,

$$\tau \ = \ \frac{Pr}{2}. \tag{5.19}$$

As the velocity is a function of r only, the only non-zero component of the strain rate tensor is

$$\dot\gamma \ = \ -\frac{dv}{dr}. \tag{5.20}$$

If we consider a time-independent non-Newtonian fluid, for which the constitutive equation is given by Eq. 5.1, we obtain

$$\frac{Pr}{2} \ = \ f\left(-\frac{dv}{dr}\right). \tag{5.21}$$

The flux Q is given by

$$Q \ = \ \int_0^R 2\pi r v\, dr \tag{5.22}$$

which after integration by parts and invoking the no-slip boundary condition that $v = 0$ when $r = R$, reduces to

$$Q \ = \ \pi \int_0^R r^2 \left(-\frac{dv}{dr}\right)\, dr. \tag{5.23}$$

Thus, knowing the precise form of the functional relation given by Eq. 5.21, the flux Q can be obtained.

In the case of the power-law model, the constitutive equation is given by Eq. 5.13. Equating Eqs. 5.13 and 5.19, we obtain

$$\dot\gamma \ = \ \left(\frac{Pr}{2\mu}\right)^{\frac{1}{n}}. \tag{5.24}$$

Using Eq. 5.20, the above equation reduces to

$$\frac{dv}{dr} = -\left(\frac{Pr}{2\mu}\right)^{\frac{1}{n}}.$$ (5.25)

Integrating this, we obtain

$$v = -\left(\frac{P}{2\mu}\right)^{\frac{1}{n}} \frac{n}{n+1} r^{1+\frac{1}{n}} + c.$$ (5.26)

The no-slip condition implies that

$$v = \left(\frac{P}{2\mu}\right)^{\frac{1}{n}} \frac{n}{n+1} (R^{1+\frac{1}{n}} - r^{1+\frac{1}{n}}).$$ (5.27)

Using Eqs. 5.23 and 5.25, the flux Q is given by

$$Q = \pi \int_0^R r^2 \left(\frac{Pr}{2\mu}\right)^{\frac{1}{n}} dr$$

$$= \left(\frac{P}{2\mu}\right)^{\frac{1}{n}} \frac{n\pi}{3n+1} R^{3+\frac{1}{n}}.$$ (5.28)

This equation is found to be good for values of n between 0.68 and 0.8 and for strain rates between 5 and 200 sec^{-1}. Note that when $n = 1$, the model reduces to the Newtonian case and the above equation becomes the Poiseuille's equation.

The power-law model is an improvement to the Poiseuille's equation by generalizing the flow to include non-Newtonian effect. The beauty of this model is its simplicity. However, there are other aspects of the flow properties of fluid which this model fails to examine, one of them being the presence of yield stress.

5.6.2 *Herschel-Bulkley Model*

This model includes the yield stress, τ_0, in its constitutive equation, which is given by Eq. 5.4 as

$$\tau = \mu\dot{\gamma}^n + \tau_0, \qquad \tau \geq \tau_0$$
$$\dot{\gamma} = 0, \qquad\qquad \tau \leq \tau_0.$$ (5.29)

Let us now determine the flux, Q, due to this model. In this case, due to the existence of a yield stress, there occurs a plug flow. The radius of the plug core region is, say, r_p (see Fig. 5.4). For $r \leq r_p$, $\tau(r) \leq \tau_0$. Thus, for the core region

$$\dot{\gamma} = 0$$

ie.,

$$\frac{dv}{dr} = 0. \tag{5.30}$$

Therefore,

$$v = \text{constant}$$

$$= v_p \text{ (say)}. \tag{5.31}$$

Hence, the velocity of flow in the core region is a constant and the velocity profile in this region is straight (see Fig. 5.4).

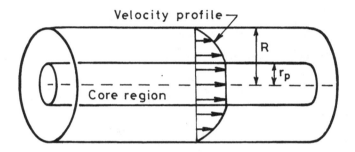

Fig. 5.4: Plug flow in the case when a yield stress exists.

Considering forces in the core region, we have

$$2\pi r_p L \tau_0 = P \times \pi r_p^2 L.$$

Thus,

$$r_p = \frac{2\tau_0}{P}, \tag{5.32}$$

or

$$\tau_0 = \frac{P r_p}{2}. \tag{5.33}$$

Outside the core region, $\tau \geq \tau_0$. The analysis in this region is similar to that in the power-law model. Using Eqs. 5.19, 5.20 and 5.29, we obtain

$$\frac{dv}{dr} = -\left(\frac{P}{2\mu}\right)^{\frac{1}{n}} (r - r_p)^{\frac{1}{n}}. \tag{5.34}$$

Integrating Eq. 5.34, we get

$$v = -\left(\frac{P}{2\mu}\right)^{\frac{1}{n}} \frac{n}{n+1} (r - r_p)^{1+\frac{1}{n}} + c. \tag{5.35}$$

Using the no-slip condition,

$$v = \left(\frac{P}{2\mu}\right)^{\frac{1}{n}} \frac{n}{n+1}\left[(R - r_p)^{1+\frac{1}{n}} - (r - r_p)^{1+\frac{1}{n}}\right]. \tag{5.36}$$

As $v = v_p$ (plug core velocity) when $r = r_p$, we have

$$v_p = \left(\frac{P}{2\mu}\right)^{\frac{1}{n}} \frac{n}{n+1}(R - r_p)^{1+\frac{1}{n}}. \tag{5.37}$$

Using the value of r_p given by Eq. 5.32 into Eqs. 5.36 and 5.37, we can determine the velocity outside the core region and in the core region respectively. Let us now determine the flux across a section of the tube.

$$
\begin{aligned}
Q &= \pi r_p^2 v_p + \int_{r_p}^{R} 2\pi r v \, dr \\[2ex]
&= \pi\left(\frac{P}{2\mu}\right)^{\frac{1}{n}} \frac{n}{n+1}\left\{r_p^2(R - r_p)^{1+\frac{1}{n}}\right. \\[2ex]
&\qquad \left. + 2\int_{r_p}^{R} r\left[(R - r_p)^{1+\frac{1}{n}} - (r - r_p)^{1+\frac{1}{n}}\right] dr\right\} \\[2ex]
&= \pi\left(\frac{P}{2\mu}\right)^{\frac{1}{n}} \frac{n}{n+1}\left\{r_p^2(R - r_p)^{1+\frac{1}{n}} + 2\left[\frac{(R^2 - r_p^2)}{2}(R - r_p)^{1+\frac{1}{n}}\right.\right. \\[2ex]
&\qquad \left.\left. -\frac{nR}{2n+1}(R - r_p)^{2+\frac{1}{n}} + \frac{(R - r_p)^{3+\frac{1}{n}}}{(2+\frac{1}{n})(3+\frac{1}{n})}\right]\right\} \\[2ex]
&= \pi\left(\frac{P}{2\mu}\right)^{\frac{1}{n}} \frac{n}{n+1}\left\{r_p^2(R - r_p)^{1+\frac{1}{n}} + 2\left[\frac{R + r_p}{2}(R - r_p)^{2+\frac{1}{n}}\right.\right. \\[2ex]
&\qquad \left.\left. -\frac{nR}{2n+1}(R - r_p)^{2+\frac{1}{n}} - \frac{(R - r_p)^{3+\frac{1}{n}}}{3+\frac{1}{n}} + \frac{(R - r_p)^{3+\frac{1}{n}}}{2+\frac{1}{n}}\right]\right\} \\[2ex]
&= \pi\left(\frac{P}{2\mu}\right)^{\frac{1}{n}} \frac{n}{n+1}\left\{r_p^2(R - r_p)^{1+\frac{1}{n}} + 2\left[\frac{R + r_p}{2}(R - r_p)^{2+\frac{1}{n}}\right.\right. \\[2ex]
&\qquad \left.\left. -\frac{n}{3n+1}(R - r_p)^{3+\frac{1}{n}} - \frac{n}{2n+1}r_p(R - r_p)^{2+\frac{1}{n}}\right]\right\} \\[2ex]
&= \pi\left(\frac{P}{2\mu}\right)^{\frac{1}{n}} \frac{n}{n+1} R^{3+\frac{1}{n}}\left[c_p^2(1 - c_p)^{1+\frac{1}{n}} + (1 + c_p)(1 - c_p)^{2+\frac{1}{n}}\right.
\end{aligned}
$$

$$-\frac{2n}{3n+1}(1-c_p)^{3+\frac{1}{n}} - \frac{2n}{2n+1}c_p(1-c_p)^{2+\frac{1}{n}}\Bigg] \qquad (5.38)$$

where

$$c_p = \frac{r_p}{R}. \qquad (5.39)$$

In the case when there is no yield stress, that is $\tau_0 = 0$, the Herschel-Bulkley model, Eq. 5.29, reduces to the power-law model, Eq. 5.2. Thus, substituting $\tau_0 = 0$ ie., $c_p = 0$ in Eq. 5.38 we obtain the value of flux for the power-law model as given in Eq. 5.28. Further, if we put $n = 1$, we obtain results for the Newtonian fluids.

Thus, the Herschel-Bulkley is a more general model compared to the power-law model. It not only considers the non-Newtonian behaviour of fluid but recognizes the existence of yield stress as well. However, it is found to agree with experimental results over a small range of shear rates.

5.6.3 *Casson Model*

The third model that will be considered is the Casson model. The constitutive equation suggested by Casson [11] is

$$\sqrt{\tau} = \sqrt{\mu\dot{\gamma}} + \sqrt{\tau_0}, \qquad \tau \geq \tau_0$$
$$\dot{\gamma} = 0, \qquad \tau \leq \tau_0. \qquad (5.40)$$

In 1958, Casson, basing his calculations on certain assumptions concerning the magnitude of inter-particle forces and disruptive stresses in chain-like floccules present in varnishes and printing inks, derived this semi-empirical equation to describe their flow behaviour. A year later, Scott Blair [58] found that this description is also suitable for the flow of blood. Since then, there has been various studies made supporting the use of the Casson equation to blood flow.

Similar to the Herschel-Bulkley model, Casson's equation includes a yield stress and therefore exhibits plug flow with radius of the core region r_p such that

$$r_p = \frac{2\tau_0}{P} \qquad (5.41)$$

and so

$$\tau_0 = \frac{Pr_p}{2}. \qquad (5.42)$$

Hence, as explained in the previous section the velocity profile in the core region, $r < r_p$, is flat. Between r_p and R (outside the core region), Casson's equation is applicable.

We will now proceed to find the velocity of flow outside the core region. Squaring the Casson equation and rearranging, we have

$$\dot{\gamma} = \frac{1}{\mu}(\tau + \tau_0 - 2\sqrt{\tau\tau_0}). \qquad (5.43)$$

Using Eq. 5.42, and knowing

$$\dot{\gamma} = -\frac{dv}{dr} \tag{5.44}$$

and

$$\tau = \frac{Pr}{2} \tag{5.45}$$

we finally obtain

$$\frac{dv}{dr} = \frac{P}{2\mu}(2\sqrt{r_p r} - r - r_p). \tag{5.46}$$

Hence,

$$v = \frac{P}{2\mu}\left(\frac{4}{3}\sqrt{r_p r^3} - \frac{r^2}{2} - r_p r + c\right). \tag{5.47}$$

Applying the no-slip condition, that is, $v = 0$ at $r = R$,

$$c = -\left(\frac{4}{3}\sqrt{r_p R^3} - \frac{R^2}{2} - r_p R\right). \tag{5.48}$$

Therefore,

$$v = \frac{P}{2\mu}\left[\frac{4}{3}\sqrt{r_p r^3} - \frac{r^2}{2} - r_p r - \left(\frac{4}{3}\sqrt{r_p R^3} - \frac{R^2}{2} - r_p R\right)\right] \tag{5.49}$$

or,

$$v = \frac{P}{4\mu}\left[(R^2 - r^2) - \frac{8}{3}\sqrt{r_p}(\sqrt{R^3} - \sqrt{r^3}) + 2r_p(R - r)\right], \quad r_p \leq r \leq R. \tag{5.50}$$

As mentioned before, the Casson model has a constant flow in the core region, giving a straight profile in this region. Let $v = v_p$ at $r = r_p$. Then

$$v_p = \frac{P}{4\mu}\left(R^2 + 2r_p R - \frac{8}{3}\sqrt{r_p R^3} - \frac{1}{3}r_p^2\right)$$

$$= \frac{P}{4\mu}(\sqrt{R} - \sqrt{r_p})^3(\sqrt{R} + \frac{1}{3}\sqrt{r_p}). \tag{5.51}$$

Equation 5.51 represents velocity distribution for all values of r between 0 and r_p, known as plug velocity.

Also,

$$v = v_p - \frac{P}{4\mu}\left(r^2 + 2r_p r - \frac{8}{3}\sqrt{r_p r^3} - \frac{1}{3}r_p^2\right). \tag{5.52}$$

The velocity profile of the flow is shown in Fig. 5.5.

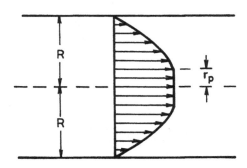

Fig. 5.5: The velocity profile of Casson flow in a long circular cylindrical pipe.

To determine the flux of the flow,

$$
\begin{aligned}
Q &= \pi r_p^2 v_p + \int_{r_p}^{R} 2\pi r v \, dr \\[2mm]
&= \pi r_p^2 v_p + 2\pi \int_{r_p}^{R} \left[v_p r - \frac{P}{4\mu}\left(r^3 + 2 r_p r^2 - \frac{8}{3}\sqrt{r_p r^5} - \frac{r_p^2 r}{3} \right) \right] dr \\[2mm]
&= \pi r_p^2 v_p + \pi v_p (R^2 - r_p^2) - \frac{P\pi}{\mu}\left[\frac{1}{8}(R^4 - r_p^4) \right. \\[2mm]
&\quad \left. + \frac{r_p}{3}(R^3 - r_p^3) - \frac{r_p^2}{12}(R^2 - r_p^2) - \frac{8}{21}\sqrt{r_p}(R^{\frac{7}{2}} - r_p^{\frac{7}{2}}) \right] \\[2mm]
&= \frac{P\pi}{\mu} R^4 \left[\left(\frac{1}{4} + \frac{1}{2}\frac{r_p}{R} - \frac{2}{3}\sqrt{\frac{r_p}{R}} - \frac{1}{12}\frac{r_p^2}{R^2} \right) - \frac{1}{8}\left(1 - \frac{r_p^4}{R^4} \right) \right. \\[2mm]
&\quad \left. - \frac{1}{3}\frac{r_p}{R}\left(1 - \frac{r_p^3}{R^3} \right) + \frac{1}{12}\frac{r_p^2}{R^2}\left(1 - \frac{r_p^2}{R^2} \right) + \frac{8}{21}\sqrt{\frac{r_p}{R}}\left(1 - \sqrt{\frac{r_p^7}{R^7}} \right) \right] \\[2mm]
&= \frac{\pi P}{8\mu} R^4 \left(1 - \frac{16}{7}c_p^{\frac{1}{2}} + \frac{4}{3}c_p - \frac{1}{21}c_p^4 \right)
\end{aligned}
\tag{5.53}
$$

where, as before

$$
c_p = \frac{r_p}{R}.
\tag{5.54}
$$

Equation 5.53 can be written as

$$Q = \frac{\pi P R^4}{8\mu} f(c_p),$$ (5.55)

where

$$\frac{Q}{Q_0} = f(c_p) = 1 - \frac{16}{7} c_p^{\frac{1}{2}} + \frac{4}{3} c_p - \frac{1}{21} c_p^4$$ (5.56)

and Q_0 denotes the flux when there is no plug flow. The variation of $f(c_p)$ with c_p is shown in Fig. 5.6. It is seen that the flux of the flow decreases rapidly with increasing c_p. For $c_p > 1$, there is no flow.

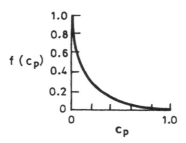

Fig. 5.6: Variation of $f(c_p)$ with c_p.

Note that when there is no yield stress, that is, $\tau_0 = 0$, the Casson equation reduces to

$$\tau = \mu \dot{\gamma}$$ (5.57)

which is the Newtonian case. Putting $\tau_0 = 0$, ie., $c_p = 0$ into Eq. 5.55 gives

$$Q = \frac{P\pi}{8\mu} R^4$$ (5.58)

which coincides with the Poiseuille's equation, Eq. 4.13. Thus, Poiseuille's model may be regarded as a particular case of Casson's model.

5.6.4 *Further Analysis of the Casson Model*

In his paper, Scott Blair [58] used the results from an experiment carried out by F.A. Glover on human blood and plotted a graph of square-root of strain rate against square-root of shear stress. The curve shows remarkable linearity, with a non-zero value for the intercept on the stress axis (see Fig. 5.7). In the same paper, Scott Blair also cited the works of Kümin, Suter, Bircher and Müller, who presented similar results for animal blood (see Fig. 5.8).

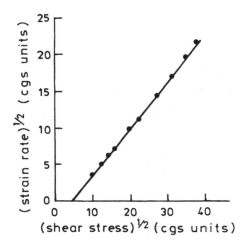

Fig. 5.7: Scott Blair's graph for healthy human blood.

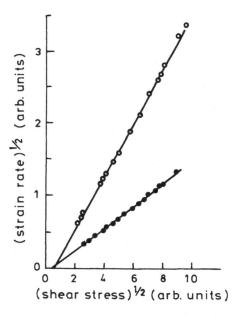

Fig. 5.8: Two samples of cow's blood (Kümin).

This linearity was further supported by the results of Merrill *et. al.* [44]. Merrill used two different methods to obtain data for suspensions of red cells in plasma of two different hematocrits. All four sets of data produced graphs which are linear and have a positive stress intercept (see Fig. 5.9).

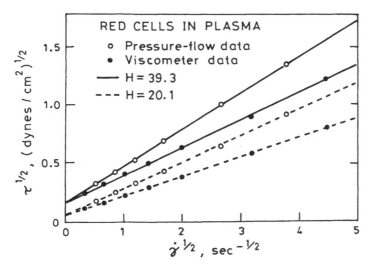

Fig. 5.9: Pressure-flow data and viscometer data on square-root coordinates.

Another reason for the use of the Casson model is that it not only agrees with experimental results, but agrees over a large range of shear rates as well. Charm and Kurland [12] demonstrated the application of Casson's equation over the range of 2 to $100,000$ sec^{-1} using a series of viscometers. The results indicated that using Casson's equation, it was possible to extrapolate blood viscometry information obtained at shear ranges of 5 to 200 sec^{-1} to shear rates of 10,000 to $100,000$ sec^{-1} with less than 5 per cent error.

5.7 Flow of Non-Newtonian Fluids in Elastic Tubes

We will now consider the effect of the elasticity of the tube wall while considering flow of non-Newtonian fluids in tubes. If the wall is elastic then the fluid velocity vector will have both axial and radial component. Further, the pressure gradient can also have both axial and radial components. Thus, it is important that we incorporate the elasticity of the tube wall into the models of non-Newtonian fluids discussed in Section 5.6.

5.7.1 *Power-Law Model Using Linear Elastic Theory*

Consider an elastic tube of length L with radius a which varies with changes in transmural pressure difference. Let the fluid enter the tube with inlet pressure p_1 and leave it with outlet pressure p_2, and the pressure outside be p_0 (see Fig. 5.10). We denote the distance along the tube from inlet end, A, by x.

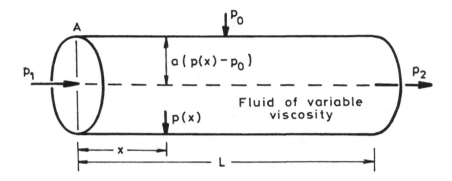

Fig. 5.10: Fluid flow in an elastic tube.

The pressure inside the tube, $p(x)$, decreases from $p(0) = p_1$, to $p(L) = p_2$. The elastic tube will expand or contract at x, depending on the pressure difference $p(x) - p_0$, known as transmural pressure difference. As such, the radius of the tube is not a constant and hence, the cross-section of the tube may deform.

It is assumed that the flux Q is related to pressure gradient $\frac{dp}{dx}$ by the relation

$$Q = -\sigma \frac{dp}{dx} \tag{5.59}$$

where σ, known as the conductivity of the tube, will be a function of the pressure difference, $p(x) - p_0$, ie.,

$$\sigma = \sigma[p(x) - p_0]. \tag{5.60}$$

Integrating Eq. 5.59 with respect to x, we have

$$Q = \frac{1}{L} \int_{p_2 - p_0}^{p_1 - p_0} \sigma(p') \, dp' \tag{5.61}$$

where, as before, we have denoted $p' = p - p_0$.

In the special case of a circular tube with radius a, a function of p', the conductivity for Newtonian fluid is given by

$$\sigma(p') = \frac{\pi}{8\mu} a^4(p'). \tag{5.62}$$

Thus, from Eq. 5.61 we obtain

$$Q = \frac{\pi}{8\mu L} \int_{p_2-p_0}^{p_1-p_0} a^4(p')\, dp' \tag{5.63}$$

which, of course, reduces to Poiseuille's formula when the radius a is a constant.

Equation 5.63 can be solved if we know the form of the function $a(p')$. If the hoop stress or tension $T(a)$ in the tube wall is known as a function of a, then $a(p')$ can be found using the equilibrium condition

$$p(x) - p_0 = \frac{T}{a}. \tag{5.64}$$

At this point it is necessary to know how the radius of a tube varies with pressure. Roach and Burton [53] determined the static pressure volume relation of a 4 cm long piece of human external iliac artery, and converted it into a tension versus length curve (see Fig. 5.11).

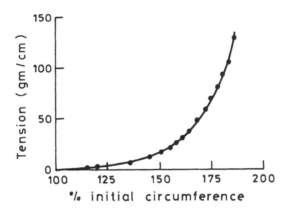

Fig. 5.11: The elastic diagram of a human iliac artery using percentage elongation.

Then, Rubinow and Keller [56] used the least squares method to fit this curve with the equation

$$T(a) = \left[t_1 \left(\frac{a}{a_0} - 1 \right) + t_2 \left(\frac{a}{a_0} - 1 \right)^5 \right] \tag{5.65}$$

where T = tension in arterial wall

$\quad a$ = radius of artery

$\quad a_o$ = 0.216 cm (unstretched radius of artery)

$\quad t_1$ = 13.0 g/cm

$\quad t_2$ = 300 g/cm.

Using the relation in Eq. 5.65, Eq. 5.64 becomes

$$p' = p(x) - p_0 = \frac{1}{a}\left[t_1\left(\frac{a}{a_0}-1\right)+t_2\left(\frac{a}{a_0}-1\right)^5\right]. \tag{5.66}$$

Differentiation gives

$$\frac{dp'}{da} = \frac{t_1}{a^2}+\frac{t_2}{a^2}\left(\frac{a}{a_0}-1\right)^4\left(\frac{4a}{a_0}+1\right). \tag{5.67}$$

Therefore, with the use of Eq. 5.67, the definite integral for flux Q appearing in Eq. 5.63 can be evaluated to give

$$Q = \frac{\pi a_0^3}{8\mu L}\left[g(\alpha_1)-g(\alpha_2)\right] \tag{5.68}$$

where

$$g(\alpha) = \frac{1}{3}t_1\alpha^3 + t_2\left(\frac{1}{2}\alpha^8 - \frac{15}{7}\alpha^7 + \frac{10}{3}\alpha^6 - 2\alpha^5 + \frac{1}{3}\alpha^3\right) \tag{5.69}$$

and α_1, α_2 are non-dimensional parameters, given by

$$\alpha_1 = \frac{a_1}{a_0}, \qquad \alpha_2 = \frac{a_2}{a_0} \tag{5.70}$$

where a_1 and a_2 are values of a, corresponding to $(p_1 - p_0)$ and $(p_2 - p_0)$ respectively. Figure 5.12 shows the variation of $g(\alpha)$ with respect to α.

The above analysis refers to a Newtonian viscous fluid. We now consider the fluid in the tube obeying the power-law given by Eq. 5.13. We have the relation between the flux and the pressure gradient, given by Eq. 5.59. Comparing this with Eq. 5.28, we conclude

$$\sigma(p') = \frac{n\pi}{\eta(3n+1)}a^{3+\frac{1}{n}}, \quad 0 < n < 1 \tag{5.71}$$

where

$$\eta = (2\mu)^{\frac{1}{n}}. \tag{5.72}$$

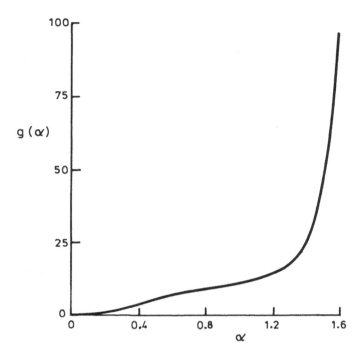

Fig. 5.12: Plot of $g(\alpha)$ vs α.

Thus, using Eqs. 5.61 and 5.71, we obtain

$$Q = \frac{n\pi}{\eta(3n+1)L} \int_{p_2-p_0}^{p_1-p_0} a^{\frac{3n+1}{n}} dp'. \tag{5.73}$$

Let us denote

$$K = \frac{n\pi}{\eta(3n+1)L}. \tag{5.74}$$

Assuming as before the tension-strain relation of the tube given by Eq. 5.65, and proceeding as before, we finally obtain

$$Q = Ka_0^s [g(\alpha_1) - g(\alpha_2)] \tag{5.75}$$

where

$$g(\alpha) = \frac{t_1\alpha^s}{s} + t_2 \left[\frac{4\alpha^{s+5}}{s+5} - \frac{15\alpha^{s+4}}{s+4} + \frac{20\alpha^{s+3}}{s+3} - \frac{10\alpha^{s+2}}{s+2} + \frac{\alpha^s}{s} \right] \tag{5.76}$$

and

$$s = \frac{2n+1}{n}, \quad 0 < n < 1. \tag{5.77}$$

At $n = 1$, Eq. 5.76 coincides with Eq. 5.69, the case for a Newtonian fluid. Since the model largely depends on the function $g(\alpha)$ for various values of the power index, n, the graph of $g(\alpha)$ for a number of values of n is shown in Fig. 5.13. It should be noted that the graph for $n = 1$ is exactly the same as that shown in Fig. 5.12.

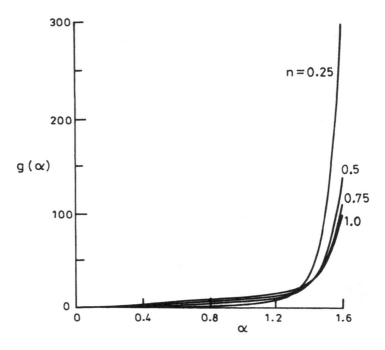

Fig. 5.13: Graph of $g(\alpha)$ for various values of n, power index.

Equation 5.75 is then the flux for the flow of a non-Newtonian fluid with variable viscosity obeying the power-law, in an elastic tube. This model may be applied to model the flow of blood in arteries. However, we need to know the value of n, the index of the power law, before this equation can be used. It should also be noted that in the equation, η is not the viscosity — η here is a constant, acting as a parameter when the "power" n is being decided. Hence, to apply the above model, prior knowledge of the viscosity of blood in the area considered is essential and this will in turn determine the appropriate values of n and μ.

The above analysis is based on the choice of a polynomial form as an approximating function for the tension-strain relationship as given by Eq. 5.65. Let us now consider the case when an exponential function is chosen as an approximating function to the experimental data. Suppose the tension $T(a)$ takes the form (Mazumdar *et. al.* [43])

$$T(\alpha) = Ae^{k\alpha} + B \qquad (5.78)$$

where

$$\alpha = \frac{a}{a_0}$$

and k, A and B are constant parameters to be determined.

Because a_0 is the initial radius, we will have an initial condition:

$$T(\alpha) = 0 \quad \text{at} \quad \alpha = 1.$$

Therefore,

$$T(\alpha) = A(e^{k\alpha} - e^k) \qquad (5.79)$$

so that A is positive. Also, from the plot of the experimental data points (Fig. 5.11), we expect k to be positive since the curve is monotonically increasing. Further, we can employ the method of least squares to find the values of A and k in Eq. 5.78. The values thus obtained are

$$A = 0.007435, \quad k = 5.262500. \qquad (5.80)$$

Hence, with these values of A and k, Eq. 5.79 becomes

$$T(\alpha) = 0.007435 \left(e^{5.2625 \frac{a}{a_0}} - e^{5.2625} \right). \qquad (5.81)$$

Proceeding as before, the value of Q can be calculated from Eq. 5.73 as

$$Q = KA \int_{a_2}^{a_1} a^{\frac{n+1}{n}} \left[\left(\frac{ka}{a_0} - 1 \right) e^{\frac{ka}{a_0}} + e^k \right] da. \qquad (5.82)$$

However, the above integral can be evaluated numerically, for instance, by using Simpson's Quadrature Formula, or the Trapezoidal Rule.

We thus obtained results using the polynomial form as well as exponential form of representing the tension-strain relationship of the vessel wall. Figure 5.14 illustrates how well the two results approximate the experimental data, given by Roach and Burton [53], and displayed in Table 5.1.

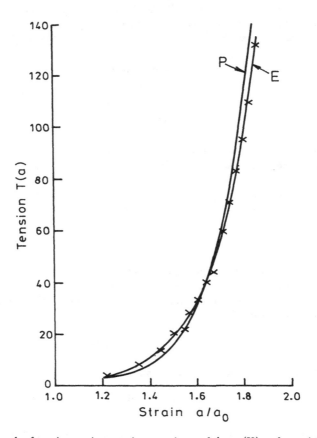

Fig. 5.14: Graph of tension against strain, experimental data, (X), polynomial approximating function, (P), and exponential approximating function, (E).

Table 5.1: Relationship between tension and strain.

T (g/cm)	4	8	13	20	22	28	33	40
a/a_0	1.22	1.35	1.45	1.50	1.55	1.57	1.60	1.64
T (g/cm)	44	60	71	83	95	109	132	
a/a_0	1.67	1.71	1.74	1.77	1.80	1.83	1.86	

It is clear from this figure that the exponential function fits the experimental data better than the polynomial function. This is almost an expected result, since the exponential is in fact an infinite series, whereas the polynomial is merely a truncated part of the series, ignoring higher order terms.

5.7.2 *Casson Model Using Linear Elastic Theory*

Consider now the flow in an elastic tube of a fluid obeying Casson's law given by Eq. 5.40. We have the relation between the flux and the pressure gradient given by Eq. 5.59. Comparing this with Eq. 5.53, we conclude

$$\sigma(p') = \frac{\pi}{8\mu} a^4 \left(1 - \frac{16}{7} c_p^{\frac{1}{2}} + \frac{4}{3} c_p - \frac{1}{21} c_p^4 \right). \tag{5.83}$$

We thus obtain the value of the flux, Q for Eq. 5.61 as

$$Q = \frac{\pi}{8\mu L} \int_{p_2 - p_0}^{p_1 - p_0} a^4 \left(1 - \frac{16}{7} c_p^{\frac{1}{2}} + \frac{4}{3} c_p - \frac{1}{21} c_p^4 \right) dp' \tag{5.84}$$

or, using the relation in Eq. 5.54, we get

$$Q = \frac{\pi}{8\mu L} \int_{p_2 - p_0}^{p_1 - p_0} \left(a^4 - \frac{16}{7} \sqrt{r_p a^7} + \frac{4}{3} r_p a^3 - \frac{1}{21} r_p^4 \right) dp'. \tag{5.85}$$

Consider now the exponential form of tension-extension for the tube wall as given in Eq. 5.81.

We have from Eq. 5.79

$$p' = p(x) - p_0 = \frac{A}{a} \left(e^{\frac{ka}{a_0}} - e^k \right). \tag{5.86}$$

Hence,

$$\frac{dp'}{da} = A \left[e^{ca} \left(\frac{c}{a} - \frac{1}{a^2} \right) + \frac{e^k}{a^2} \right] \tag{5.87}$$

where

$$c = \frac{k}{a_0}. \tag{5.88}$$

Equation 5.85 now takes the form

$$
\begin{aligned}
Q &= \frac{A\pi}{8\mu L} \int_{a_2}^{a_1} \left(a^4 - \frac{16}{7} \sqrt{r_p a^7} + \frac{4}{3} r_p a^3 - \frac{1}{21} r_p^4 \right) \left(e^{ca} \left(\frac{c}{a} - \frac{1}{a^2} \right) + \frac{e^k}{a^2} \right) da \\
&= \frac{A\pi}{\mu L} \int_{a_2}^{a_1} \left[e^k \left(\frac{a^2}{8} + \frac{r_p}{6} a - \frac{2}{7} \sqrt{r_p a^3} - \frac{r_p^4}{168} \frac{1}{a^2} \right) + e^{ca} \left(\frac{c}{8} a^3 + \frac{r_p c}{6} a^2 \right. \right. \\
&\quad \left. \left. - \frac{2c}{7} \sqrt{r_p a^5} - \frac{r_p^4 c}{168} \frac{1}{a} - \frac{a^2}{8} - \frac{r_p}{6} a + \frac{2}{7} \sqrt{r_p a^3} + \frac{r_p^4}{168} \frac{1}{a^2} \right) \right] da
\end{aligned} \tag{5.89}
$$

where $A = 0.007435$
$ k = 5.2625$
$ a_0 = 0.216$ cm
$ c = \frac{k}{a_0}$.

The above integral, after numerical evaluation, gives the value of flux for Casson's model in an elastic tube.

The results in this chapter give complete information on the laminar flow in circular cylindrical tubes.

Chapter 6

MODELS FOR OTHER FLOWS

6.1 Introduction

In this chapter some applications of the basic equations derived earlier are demonstrated and a number of models of other flows are discussed. In particular, a model of oxygen diffusion from blood vessel to tissue, as proposed by Krogh [36], a model for fluid movement in renal tubules in kidneys, and diffusion in artificial kidneys, a model for the measurement of the extravascular space by the indicator dilution method, and finally a model for the peristaltic flow, which is the motion of a fluid in an elastic tube when a wave of contraction and expansion travels along the wall of the tube, are discussed in detail.

A comprehensive study of the dynamics of flows of other biofluids is beyond the scope of this book. There are many books and papers on this subject, amongst which the following are worth mentioning; Fung [19,20,21], Kapur [33] and Lighthill [39,40]. We will emphasize in this chapter a quantitative way of assessing certain aspects of these models.

One of the basic mechanisms in biological systems is that of the process of diffusion in the human system. Accordingly, in the next section we discuss this process in order to explain the diffusion of oxygen from blood in capillaries to the surrounding tissues.

105

6.2 The Krogh Model of Oxygen Diffusion From Blood Vessel to Tissue

In this model, the capillary blood vessel is represented by a straight circular cylindrical tube and the tissue is represented by a stationary concentric cylindrical tube (see Fig. 6.1). A large number of identical tubes are closely packed to represent the tissue of an organ (Fig. 6.2).

The model is used to study the transport of molecular oxygen from the blood plasma to the skeletal muscle tissues or lungs or brain, across the capillary walls. The model will give an indication of the effect of the velocity of blood flow in capillaries, axial and radial diffusion of oxygen in blood, oxygen diffusivity in tissues, and permeability of capillary walls.

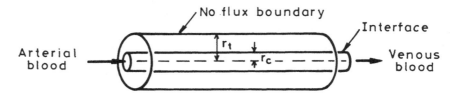

Fig. 6.1: Krogh tissue cylinder.

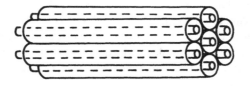

Fig. 6.2: Closely packed identical tubes.

The model hypothesizes that, in a given portion of tissue, all capillaries and surrounding tissues are of equal diameter and are homogeneously dispersed in the tissue. In the capillary vessel, oxygen transport takes place both by convection and by diffusion. In the tissue there is diffusion of oxygen, as well as consumption of oxygen by the tissue cells.

Since diffusion is the main mechanism, we will first introduce the basic equations of the process of diffusion. For more detail see Mazumdar [42]. Let $c(\mathbf{x}, t)$ be the concentration of a solute or the amount of solute per unit volume at the point $\mathbf{x}$ at time t. Due to the concentration gradient, there is a flow of solute given by the flux vector or the current density vector $\mathbf{J}$. This vector consists of two parts: diffusive

$\mathbf{J}_{diff}$ and convective $\mathbf{J}_{conv}$. The former obeys Fick's law, according to which the flux is proportional to the concentration gradient. The latter is proportional to the product of velocity and concentration. Hence, we have

$$\mathbf{J}_{diff} = -D\nabla c, \qquad (6.1)$$

$$\mathbf{J}_{conv} = c\mathbf{q} \qquad (6.2)$$

where $\mathbf{q}$ (with components u, v, w) is the flow velocity vector and D is the coefficient of diffusion. Combining these two current densities we have the total flux vector $\mathbf{J} = (J_x, J_y, J_z)$ given by

$$J_x = -D_x \frac{\partial c}{\partial x} + uc, \quad J_y = -D_y \frac{\partial c}{\partial y} + vc, \quad J_z = -D_z \frac{\partial c}{\partial z} + wc, \qquad (6.3)$$

where D_x, D_y, D_z are the coefficients of diffusion in the x, y, z directions respectively.

We will now apply the conservation principle to the transport of solute. The conservation principle is a fundamental principle of nature and all natural phenomena must, whether it is physical, biological or chemical, conform with this principle.

Consider a very small cubical volume of solution (as shown in Fig. 6.3).

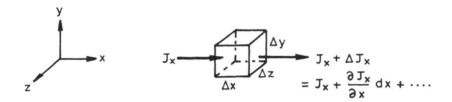

Fig. 6.3: Conservation of solute transport into and out of a small volume element.

Conservation of solute transport into and out of this volume element states that, within a given infinitesimal volume element of a solution in which solute currents exist, whatever the cause of the currents may be, the rate at which matter accumulates or disappears within the region is equal to the net flux across the surface bounding the infinitesimal region. This is stated by the following expression:

$$\left\{ J_x dy dz - \left(J_x + \frac{\partial J_x}{\partial x} dx \right) dy dx \right\} + \left\{ J_y dx dz - \left(J_y + \frac{\partial J_y}{\partial y} dy \right) dx dz \right\}$$

$$+ \left\{ J_z dx dy - \left(J_z + \frac{\partial J_z}{\partial z} dz \right) dx dy \right\} = \left(g + \frac{\partial c}{\partial t} \right) dx dy dz \qquad (6.4)$$

where the LHS of the above equation is the net diffusive flux of solute across the surface bounding the volume element and the RHS is the increment per unit time of the concentration of the infinitesimal volume element. Here, we assumed that due to chemical reaction in the volume element, the metabolite is consumed at a rate g moles per unit volume per unit time. By simplifying Eq. 6.4 we obtain

$$g + \frac{\partial c}{\partial t} + \frac{\partial J_x}{\partial x} + \frac{\partial J_y}{\partial y} + \frac{\partial J_z}{\partial z} = 0. \tag{6.5}$$

Substituting Eqs. 6.3 into Eq. 6.5, we obtain

$$g + \frac{\partial c}{\partial t} = \frac{\partial}{\partial x}\left(D_x \frac{\partial c}{\partial x}\right) + \frac{\partial}{\partial y}\left(D_y \frac{\partial c}{\partial y}\right) + \frac{\partial}{\partial z}\left(D_z \frac{\partial c}{\partial z}\right)$$
$$- c\left(\frac{\partial u}{\partial x} + \frac{\partial v}{\partial y} + \frac{\partial w}{\partial z}\right) - u\frac{\partial c}{\partial x} - v\frac{\partial c}{\partial y} - w\frac{\partial c}{\partial z} \tag{6.6}$$

which is the basic equation for diffusion. If the fluid is incompressible, then

$$\frac{\partial u}{\partial x} + \frac{\partial v}{\partial y} + \frac{\partial w}{\partial z} = 0. \tag{6.7}$$

If, in addition, the coefficient of diffusion D is a constant, then Eq. 6.6 is reduced to

$$\frac{Dc}{Dt} \equiv \frac{\partial c}{\partial t} + u\frac{\partial c}{\partial x} + v\frac{\partial c}{\partial y} + w\frac{\partial c}{\partial z}$$
$$= D\left(\frac{\partial^2 c}{\partial x^2} + \frac{\partial^2 c}{\partial y^2} + \frac{\partial^2 c}{\partial z^2}\right) - g. \tag{6.8}$$

We will now discuss this equation in the capillary region as well as in the tissue region.

6.2.1 *Capillary Blood Vessel Region*

In the capillaries, assume the distribution of oxygen is axisymmetric. Let $v(r, t)$ denote the velocity of blood in the fully developed flow, $c(r, z, t)$ be the concentration of oxygen, $d(c)$ be the rate of generation of oxygen per unit volume and r_c be the radius of a capillary (see Fig. 6.1). Hence, after using cylindrical polar coordinates, Eq. 6.8 gives

$$\frac{Dc}{Dt} = D_b\left(\frac{\partial^2 c}{\partial r^2} + \frac{1}{r}\frac{\partial c}{\partial r} + \frac{\partial^2 c}{\partial z^2}\right) + d(c), \qquad 0 < r < r_c \tag{6.9}$$

where D_b denotes the overall coefficient of diffusion of oxygen in the blood.

For large capillaries, Eq. 6.9 is satisfactory, but for small capillaries, it is usually preferable to use two diffusivity coefficients, ie., D_r in the radial direction and D_b in the axial direction.

6.2.2 *Tissue Region*

In this region, let D_t denote the coefficient of diffusion for oxygen. The velocity of motion of the tissue is zero and hence, there is no convective term. In this region, the basic diffusion equation (6.8) becomes

$$\frac{\partial c}{\partial t} = D_t \left(\frac{\partial^2 c}{\partial r^2} + \frac{1}{r}\frac{\partial c}{\partial r} + \frac{\partial^2 c}{\partial z^2} \right) - g(c), \quad r_c \le r \le r_t. \tag{6.10}$$

It is assumed that the oxygen consumption rate $g(c)$ appearing in the above equation follows Michaelis-Menton kinetics so that

$$g(c) = \frac{\lambda c}{\mu + c} \tag{6.11}$$

where λ and μ are constants.

If c is large in Eq. 6.11, $g(c)$ becomes a constant, say g_0, whereas if c is small, $g(c)$ will be of the form $\lambda_1 c$, where λ_1 is a constant.

6.2.3 *Boundary Conditions*

The boundary conditions for this situation are as follows:

(i) Due to the symmetry of the capillary flow about the axis, the radial component of the concentration gradient vanishes, ie.,

$$\frac{\partial c}{\partial r} = 0 \quad \text{at} \quad r = 0. \tag{6.12a}$$

(ii) There is no flux at the surface of the tissue, ie.,

$$\frac{\partial c}{\partial r} = 0 \quad \text{at} \quad r = r_t. \tag{6.12b}$$

(iii) The continuity condition at the interface requires that the flux of oxygen must be continuous at the blood-tissue interface, so that

$$\left(D_b \frac{\partial c}{\partial r} \right)_{\text{blood}} = \left(D_t \frac{\partial c}{\partial r} \right)_{\text{tissue}} \quad \text{at} \quad r = r_c. \tag{6.12c}$$

6.2.4 *Krogh's Steady-State Model*

Krogh [36] obtained the steady-state solution for the tissue region, assuming that $g(c)$ in Eq. 6.10 is a constant g_0 and there is no axial diffusion.

Under these assumptions, Eq. 6.10 reduces, in steady state, to

$$\frac{\partial^2 c}{\partial r^2} + \frac{1}{r}\frac{\partial c}{\partial r} = \frac{g_0}{D_t}. \tag{6.13}$$

The solution to this equation, subjected to the boundary condition 6.12b, together with the condition that $c = c_0$ at $r = r_c$, is

$$c = c_0 + \frac{g_0}{D_t}\left(\frac{r^2 - r_c^2}{4} - \frac{r_t^2}{2}\ln\frac{r}{r_c}\right). \tag{6.14}$$

The above equation can be written as

$$\frac{c_0 - c}{g_0 r_c^2 D_t^{-1}} = \frac{1}{2}R^2\ln x - \frac{1}{4}(x^2 - 1) \tag{6.15}$$

where

$$R = \frac{r_t}{r_c}, \quad x = \frac{r}{r_c}, \quad 1 \le x \le R. \tag{6.16}$$

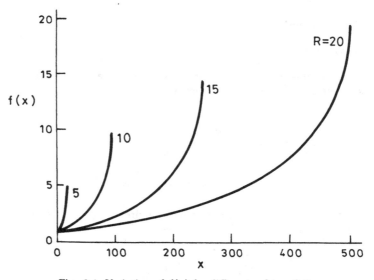

Fig. 6.4: Variation of $f(x)$ for different values of R.

Denoting

$$\frac{1}{2}R^2 \ln x - \frac{1}{4}(x^2 - 1) = f(x) \qquad (6.17)$$

the variation of $f(x)$ with x for various values of R is shown in Fig. 6.4.

6.2.5 *Blum's Steady-State Model*

Blum [9] relaxed some of the conditions imposed in Krogh's model and obtained the steady-state solution for the tissue region. The basic equations and the boundary condition for this model are as follows:

$$D_t \frac{1}{r}\frac{\partial}{\partial r}\left(r\frac{\partial c}{\partial r}\right) = kc, \qquad r_c \le r \le r_t, \qquad (6.18)$$

$$\frac{\partial c}{\partial r} = 0 \quad \text{at} \quad r = r_t, \qquad (6.19)$$

$$\bar{v}\frac{\partial \bar{c}}{\partial z} = \frac{2}{r_c}\left(D_t\frac{\partial c}{\partial r}\right)_{r=r_c}, \qquad (6.20)$$

$$\left(D_t\frac{\partial c}{\partial r}\right)_{r=r_c} = \left(D_b\frac{\partial c}{\partial r}\right)_{r=r_c} = P(c_0 - \bar{c}), \qquad (6.21)$$

$$\bar{c} = c_{in} \quad \text{at} \quad z = 0, \quad 0 \le r \le r_c, \qquad (6.22)$$

where $\bar{v}$ is the average velocity of blood in the capillary, c_{in} is the concentration of oxygen at the capillary inlet, and $\bar{c}(z)$ is the average concentration, given by

$$\bar{c}(z) = \frac{1}{\pi r_c^2}\int_0^{r_c} c(r, z)\, 2\pi r\, dr. \qquad (6.23)$$

Clearly, Blum's model considered first-order kinetics when $g(c) = kc$, and considered the axial variation of oxygen concentration in the capillary blood vessel.

We notice that the basic equation (6.18) is the modified Bessel equation

$$\frac{\partial^2 c}{\partial r^2} + \frac{1}{r}\frac{\partial c}{\partial r} - \frac{k}{D_t}c = 0 \qquad (6.24)$$

which reduces to

$$\frac{\partial^2 c}{\partial r^2} + \frac{1}{r}\frac{\partial c}{\partial r} - \frac{\lambda^2}{r_c^2}c = 0 \qquad (6.25)$$

if we substitute

$$\lambda^2 = \frac{kr_c^2}{D_t}. \qquad (6.26)$$

The solution of Eq. 6.25 is given by

$$c(r, z) = A(z)I_o\left(\frac{\lambda r}{r_c}\right) + B(z)K_0\left(\frac{\lambda r}{r_c}\right) \tag{6.27}$$

where I_0 and K_0 are the modified Bessel functions of zero order, and $A(z), B(z)$ are arbitrary functions of z.

Using the boundary condition (6.19), we get

$$A(z)I_1\left(\frac{\lambda r_t}{r_c}\right) - B(z)K_1\left(\frac{\lambda r_t}{r_c}\right) = 0. \tag{6.28}$$

Using this result in Eq. 6.27 we obtain

$$c(r, z) = A(z)\left[I_0\left(\frac{\lambda r}{r_c}\right) + \frac{I_1\left(\frac{\lambda r_t}{r_c}\right)}{K_1\left(\frac{\lambda r_t}{r_c}\right)}K_0\left(\frac{\lambda r}{r_c}\right)\right]. \tag{6.29}$$

If we put $r = r_c$, we obtain

$$c(r_c, z) = A(z)\left[I_0(\lambda) + \frac{I_1\left(\frac{\lambda r_t}{r_c}\right)}{K_1\left(\frac{\lambda r_t}{r_c}\right)}K_0(\lambda)\right] = c_0(z) \quad \text{(say)}. \tag{6.30}$$

So substituting the value of $A(z)$ obtained from Eq. 6.30 into Eq. 6.29, we get

$$c(r, z) = c_0(z)F\left(\frac{r}{r_c}, \frac{r_t}{r_c}, \lambda\right) \tag{6.31}$$

where

$$F\left(\frac{r}{r_c}, \frac{r_t}{r_c}, \lambda\right) = \frac{I_0\left(\frac{\lambda r}{r_c}\right)K_1\left(\frac{\lambda r_t}{r_c}\right) + I_1\left(\frac{\lambda r_t}{r_c}\right)K_0\left(\frac{\lambda r}{r_c}\right)}{I_0(\lambda)K_1\left(\frac{\lambda r_t}{r_c}\right) + I_1\left(\frac{\lambda r_t}{r_c}\right)K_0(\lambda)} \tag{6.32}$$

and

$$F\left(\frac{r}{r_c}, \frac{r_t}{r_c}, \lambda\right)_{r=r_c} = 1. \tag{6.33}$$

Differentiating Eq. 6.31 and putting $r = r_c$ we obtain

$$\left(\frac{\partial c}{\partial r}\right)_{r=r_c} = c_0(z)\frac{\lambda}{r_c}\left[\frac{I_1(\lambda)K_1\left(\frac{\lambda r_t}{r_c}\right) - I_1\left(\frac{\lambda r_t}{r_c}\right)K_1(\lambda)}{I_0(\lambda)K_1\left(\frac{\lambda r_t}{r_c}\right) + I_1\left(\frac{\lambda r_t}{r_c}\right)K_0(\lambda)}\right] \tag{6.34}$$

$$= -c_0\left(\frac{\lambda}{r_c}\right)B \quad \text{(say)} \tag{6.35}$$

where the quantity in Bessel functions in Eq. 6.34 is denoted by B.

Using Blum's boundary condition (6.21) we get

$$D_t c_0 \left(\frac{\lambda}{r_c}\right) B + P(c_0 - \bar{c}) = 0 \tag{6.36}$$

or

$$\bar{c}(z) = c_0(z)(1 + \beta B) \tag{6.37}$$

where

$$\beta = \frac{\lambda D_t}{Pr_c} = \frac{(kD_t)^{\frac{1}{2}}}{P}. \tag{6.38}$$

Thus, from Eq. 6.31 we finally obtain Blum's steady-state solution in the tissue region

$$c(r, z) = \frac{\bar{c}(z)}{1 + \beta B} F\left(\frac{r}{r_c}, \frac{r_t}{r_c}, \lambda\right). \tag{6.39}$$

Clearly, when $P \to \infty$, the above solution gives

$$c_0(z) = \bar{c}. \tag{6.40}$$

6.3 Fluid Flow in Kidneys

6.3.1 *Introduction*

One of the main functions of kidneys is to maintain the chemical balance of blood by excreting the waste products such as urea, creatine and uric acid in the blood stream. If the urea is not removed, a disease known as *uremia* occurs which may sometimes be fatal. Sometimes, the impure blood is taken out of the body. The urea is then removed from it with the help of an external device known as a *haemodialyser*, and the purified blood is returned to the body. This process is known as the *dialysis of blood*.

An important diffusion process takes place in renal tubules in kidneys, in the form of diffusion of urea through the semi-permeable membrane of a dialyser. Considering the variation of urea concentration in the dialyzate, the problem reduces to solving two partial differential equations for the blood and the dialyzate regions.

Also, an important fluid flow takes place in the functional unit of the kidney known as the *nephron* or *renal tubule*. Each kidney has about a million of these tubules. Here, because of reabsorption of water and other low molecular weight substances along the walls of the renal tube we have both radial and axial velocity of flow.

6.3.2 *Diffusion Process in the Haemodialyser*

A haemodialyser, sometimes called an *artificial kidney*, is regarded as a concentric circular duct (Fig. 6.5). Blood flows on the inside, and outside, the fluid called dialyzate which is a solution of some chemicals in water, flows. The wall of the inner cylinder is a semi-permeable membrane that allows urea to diffuse through it. During the process of flow in the duct, blood loses urea which permeates through the duct wall into the dialyzate. By maintaining a continuous supply of fresh dialyzate, the concentration of urea in the dialyzate is maintained lower than that of the blood.

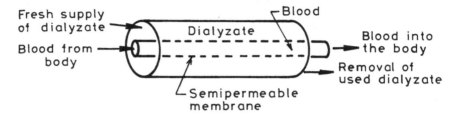

Fig. 6.5: Circular-duct haemodialyzer.

The basic governing equation is the diffusion equation given by Eq. 6.8 in cylindrical coordinates

$$D\left(\frac{\partial^2 c}{\partial r^2} + \frac{1}{r}\frac{\partial c}{\partial r} + \frac{\partial^2 c}{\partial z^2}\right) = \frac{\partial c}{\partial t} + v\frac{\partial c}{\partial z} \tag{6.41}$$

where $c(r, z, t)$ is the concentration of urea in the blood and $v(r, t)$ is the velocity in the fully-developed flow. If we assume steady-state laminar flow in a straight duct and neglect the longitudinal diffusion term, we then have

$$D\left(\frac{\partial^2 c}{\partial r^2} + \frac{1}{r}\frac{\partial c}{\partial r}\right) = v_m\left(1 - \frac{r^2}{R^2}\right)\frac{\partial c}{\partial z} \tag{6.42}$$

where v_m is the maximum velocity in the duct and R is the radius of the duct.

The boundary conditions for this flow are as follows:

(i) $\dfrac{\partial c}{\partial r} = 0$ at $r = 0$, $\hspace{6cm}$ (6.43a)

(ii) $c = c_{in}$ at $z = 0$, $0 \le r \le R$, $\hspace{4.2cm}$ (6.43b)

(iii) $-D\dfrac{\partial c}{\partial r} = P(c - c_d)$ at $r = R$, $\hspace{4cm}$ (6.43c)

where the last condition follows on the assumption of constant permeability P and constant concentration c_d in the dialyzate.

Introducing non-dimensional variables

$$\bar{c} = \frac{c - c_d}{c_{in} - c_d}, \quad \bar{r} = \frac{r}{R}, \quad P_e = \frac{v_m R}{D}, \quad \bar{z} = \frac{z}{R P_e} \tag{6.44}$$

where P_e is the dimensionless Peclet number, Eqs. 6.42 and 6.43 finally reduce to

$$\frac{\partial^2 \bar{c}}{\partial \bar{r}^2} + \frac{1}{\bar{r}} \frac{\partial \bar{c}}{\partial \bar{r}} = (1 - \bar{r}^2) \frac{\partial \bar{c}}{\partial \bar{z}}, \tag{6.45}$$

$$\frac{\partial \bar{c}}{\partial \bar{r}} = 0 \quad \text{at} \quad \bar{r} = 0, \tag{6.46a}$$

$$\bar{c} = 1 \quad \text{at} \quad \bar{z} = 0, \quad 0 \le \bar{r} \le 1, \tag{6.46b}$$

$$\frac{\partial \bar{c}}{\partial \bar{r}} + Sh_w \bar{c} = 0 \quad \text{at} \quad \bar{r} = 1, \quad \bar{z} > 0, \tag{6.46c}$$

where Sh_w is called the *Sherwood wall number*, given by

$$Sh_w = \frac{PR}{D}. \tag{6.47}$$

The solution to Eq. 6.45 with the boundary conditions (6.46) can be obtained in a variety of ways. If we use the method of separation of variables, assume the solution is of the form (Kapur [33])

$$\bar{c}(\bar{r}, \bar{z}) = \sum_{n=0}^{\infty} A_n R_n(\bar{r}) \exp\left(-\lambda_n^2 \bar{z}\right) \tag{6.48}$$

where the λ_n's are the eigenvalues, R_n's are eigenfunctions and coefficient A_n's are to be determined using the orthogonality properties of eigenfunctions.

Substituting Eq. 6.48 into Eq. 6.45 yields

$$\frac{d^2 R_n}{d\bar{r}^2} + \frac{1}{\bar{r}} \frac{dR_n}{d\bar{r}} + \lambda_n^2 (1 - \bar{r}^2) R_n = 0 \tag{6.49}$$

with boundary conditions

$$\frac{dR_n}{d\bar{r}} = 0 \quad \text{at} \quad \bar{r} = 0, \tag{6.50a}$$

$$\sum_{n=0}^{\infty} A_n R_n = 1 \quad \text{for} \quad 0 \le \bar{r} \le 1, \tag{6.50b}$$

$$\frac{dR_n}{d\bar{r}} + Sh_w R_n = 0 \quad \text{at} \quad \bar{r} = 1. \tag{6.50c}$$

The solution $R_n(\bar{r})$ to Eq. 6.49 satisfying the boundary conditions in Eqs. 6.49a-c can thus be obtained in the form of a power series in $\bar{r}$, which when substituted into

Eq. 6.48 gives the required concentration of urea in the blood in a haemodialyser. The same analysis can be made if instead of a circular-duct dialyser, we consider a parallel-plate dialyser.

6.3.3 *Flow in the Renal Tubule*

The renal tubule or nephron is the functional unit of the kidney. Each kidney has about one million of these tubules. The main function of these tubules is to filter the blood coming from the renal artery. During the course of filtration this filtrate is subsequently reabsorbed by the collecting ducts.

Due to this reabsorption, both radial and axial components of the velocity of flow exist (Fig. 6.6). Hence, if we neglect the inertial terms since the Reynolds number in this case is very much less than one (about 10^{-3}), we have the following basic equations corresponding to Eqs. 1.44-1.49 discussed in Chapter 1.

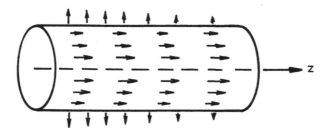

Fig. 6.6: Flow in a renal tubule.

$$\frac{1}{r}\frac{\partial}{\partial r}(rv_r) + \frac{\partial v_z}{\partial z} = 0, \tag{6.51}$$

$$\frac{\partial^2 v_r}{\partial r^2} + \frac{1}{r}\frac{\partial v_r}{\partial r} - \frac{v_r}{r^2} + \frac{\partial^2 v_r}{\partial z^2} = \frac{1}{\mu}\frac{\partial p}{\partial r}, \tag{6.52}$$

$$\frac{\partial^2 v_z}{\partial r^2} + \frac{1}{r}\frac{\partial v_z}{\partial r} + \frac{\partial^2 v_z}{\partial z^2} = \frac{1}{\mu}\frac{\partial p}{\partial z}. \tag{6.53}$$

The boundary conditions for this flow are as follows

(i) $\dfrac{\partial v_z}{\partial r} = 0, \quad v_r = 0 \quad$ at $\quad r = 0,$ \hfill (6.54a)

(ii) $v_z = 0, \quad v_r = \phi(z) \quad$ at $\quad r = R.$ \hfill (6.54b)

Eliminating p between Eqs. 6.52 and 6.53 and introducing the stream function ψ, such that

$$v_z = \frac{1}{r}\frac{\partial \psi}{\partial r}, \qquad v_r = -\frac{1}{r}\frac{\partial \psi}{\partial z}, \tag{6.55}$$

we finally obtain, as shown in Eq. 1.48

$$\nabla_1^4 \psi = 0 \tag{6.56}$$

where

$$\nabla_1^2 \equiv \frac{\partial^2}{\partial r^2} - \frac{1}{r}\frac{\partial}{\partial r} + \frac{\partial^2}{\partial z^2}. \tag{6.57}$$

In order to solve Eq. 6.56, let us try a solution of the form

$$\psi(r, z) = f(r)\left(a_0 z + \frac{1}{2}a_1 z^2\right) + g(r). \tag{6.58}$$

The above form of the solution is based on the assumption that the function $\phi(z)$, which is the value of v_r at $r = R$, is a linear function in z.

We thus have, using Eq. 6.55

$$v_r = -\frac{1}{r}f(r)(a_0 + a_1 z), \tag{6.59}$$

$$v_z = \frac{1}{r}f'(r)\left(a_0 z + \frac{1}{2}a_1 z^2\right) + g'(r) \tag{6.60}$$

and Eq. 6.56 gives

$$\left(a_0 z + \frac{1}{2}a_1 z^2\right)D^2 f(r) + D^2 g(r) + 2a_1 D f(r) = 0 \tag{6.61}$$

where the operator D is given by

$$D \equiv \frac{d^2}{dr^2} - \frac{1}{r}\frac{d}{dr}. \tag{6.62}$$

We thus have,

$$D^2 f(r) = 0, \tag{6.63}$$

$$D^2 g(r) + 2a_1 D f(r) = 0. \tag{6.64}$$

Solving Eqs. 6.63 and 6.64 using the required boundary conditions, we finally obtain

$$f(r) = \frac{2r^2}{R} - \frac{r^4}{R^3} \tag{6.65}$$

and

$$g(r) = \left(\frac{a_1 R}{12} - \frac{Q_0}{\pi R^2}\right)r^2 + \left(\frac{Q_0}{2\pi R^4} - \frac{a_1}{6R}\right)r^4 + \frac{a_1}{12R^3}r^6 \tag{6.66}$$

where Q_0 is the value of flux at the entrance $(z = 0)$, ie.,

$$Q_0 = \int_0^R 2\pi r v_z \, dr. \tag{6.67}$$

We can thus calculate the velocity components v_r and v_z, and the pressure $p(r, z)$. It can be shown that in this case

$$v_z = \frac{2Q(z)}{\pi R^2}\left[\left(1 - \frac{r^2}{R^2}\right) - \frac{a_1 R}{2}\left(\frac{1}{3} - \frac{r^2}{R^2}\right)\right] \tag{6.68}$$

in contrast to the expression for Poiseuille's flow

$$v_z = \frac{2Q}{\pi R^2}\left(1 - \frac{r^2}{R^2}\right) \tag{6.69}$$

and

$$p(r, z) = p_0 - \frac{4\mu}{R^3}(a_0 + a_1 z)r^2 - \mu\left(\frac{4a_1}{3R} + \frac{8\overline{Q}(z)}{\pi R^4}\right)z \tag{6.70}$$

where

$$Q(z) = Q_0 - \pi R(2a_0 z + a_1 z^2) \tag{6.71}$$

and

$$\overline{Q}(z) = \int_0^z Q(z) \, dz. \tag{6.72}$$

The above results are obtained assuming a particular form of the stream function given by Eq. 6.58. Let us now consider another form of the stream function, given by

$$\psi(r, z) = f(r)e^{-\alpha z} + g(r) \tag{6.73}$$

where we have assumed that the radial velocity at the wall of the cylinder decreases exponentially with z.

Proceeding as before, we obtain

$$D_1^2 f(r) = 0, \tag{6.74}$$

$$D^2 g(r) = 0 \tag{6.75}$$

where

$$D_1 \equiv \frac{d^2}{dr^2} - \frac{1}{r}\frac{d}{dr} + \alpha^2. \tag{6.76}$$

Solving Eqs. 6.74 and 6.75 we get the functions $f(r)$ and $g(r)$, then using Eq. 6.55 we get velocity components v_r and v_z, and the pressure $p(r, z)$. It can be shown that in this case

$$v_r = \frac{J_1(\alpha R)J_1(\alpha r) - \frac{r}{R}J_0(\alpha R)J_2(\alpha r)}{J_1^2(\alpha R) - J_0(\alpha R)J_2(\alpha R)}v_0 e^{-\alpha z}, \tag{6.77}$$

$$v_z = \frac{J_1(\alpha R)J_0(\alpha r) - \frac{r}{R}J_0(\alpha R)J_1(\alpha r)}{J_1^2(\alpha R) - J_2(\alpha R)J_0(\alpha R)}v_0 e^{-\alpha z} + \left(\frac{2Q_0}{\pi R^2} - \frac{4v_0}{\alpha R}\right)\left(1 - \frac{r^2}{R^2}\right) \quad (6.78)$$

and

$$p(r,z) = p_0 + \frac{2v_0}{R}\left(1 - e^{-\alpha z}\right)\frac{J_0(\alpha R)}{J_0(\alpha R)J_2(\alpha R) - J_1^2(\alpha R)} - \frac{4z}{R^2}\left(\frac{2Q_0}{\pi R^2} - \frac{4v_0}{\alpha R}\right) \quad (6.79)$$

where we have assumed that the function $\phi(z)$, which is the value of v_r at $r = R$ is given by

$$\phi(z) = v_0 e^{-\alpha z}. \quad (6.80)$$

6.4 Flow Measurement by Indicator Dilution Method

6.4.1 *Introduction*

An indicator or a tracer is a substance which in some way labels a moving fluid so that the movement of the indicator is representative to that of the fluid which it labels. An indicator can be introduced into the living body and is not metabolized. The presence of an indicator does not alter any properties of the system and it can easily be recovered from the system. The distribution of an indicator in the system is considered to be a linear function of the input.

6.4.2 *Measurement of Flow*

Let us consider the case of a vascular system which has only a single arterial inlet and a single venous outlet. We shall assume that the indicator is a dye. Suppose an amount m of an indicator is injected into the blood stream on the venous side of the left ventricle. We assume that the ventricle behaves as a reservoir having the same rate of influx as that of outflux. This assumption obviously neglects the periodic nature of blood flow. It is further assumed that the blood in the veins containing the full amount of dye will enter the left ventricle and then proceed to the aorta. The concentration of dye in the aorta can be measured.

The method is based on the simple concept that a known amount of indicator is injected into the blood stream and the rate of flow is measured by the rate of dilution of the indicator. It is, therefore, based on the law of conservation of mass. A similar method, known as Allen's method, is used in fluid engineering for determining the volume of water in a conduit.

Although a variety of indicators, such as sodium, iodine, nitrous oxide, carbon monoxide, acetylene, etc. have been used, the most frequently used indicator is cardiogree dye or indocyanine green.

Let $c(t)$ be the concentration of dye in the left ventricle, V be the volume of blood flow in the left ventricle and k be the *Cardiac Output* (CO), the flux of blood through the left ventricle or the steady rate at which blood is ejected by the heart. The cardiac output is an important parameter in determining the condition of the heart with reference to its proper functioning.

We assume that all the dye moves with the blood from the left ventricle to the aorta, and no amount of dye can escape through any of the upstream arteries.

Let the time be measured from the instant when the dye is injected into a systemetic vein and t_0 be the time when the dye enters into the ventricle. The initial amount of dye that is injected and observed into the ventricle is

$$m = Vc(t_0). \tag{6.81}$$

We therefore have the basic equation for the concentration of dye from one-compartment model analysis (Mazumdar [42])

$$\frac{d}{dr}(Vc) = -kc, \quad t > t_0, \tag{6.82}$$

whose solution is given by

$$c(t) = \frac{m}{V}e^{-\frac{k}{V}(t-t_0)}, \quad t > t_0. \tag{6.83}$$

It is interesting to note that the actual experimental curve which is obtained by Zierler [75] shows the following facts: The time that the dye takes to reach the heart is approximately 5 secs. After this time delay, the concentration of dye initially increases linearly to a maximum value and then the curve behaves exponentially. The initial linear increase may be accounted for by reasons such as the time taken for the injection, the process of diffusion that takes place after the injection and the transport phenomenon for the dye to reach the left ventricle. However, if the experimental curve, say $\bar{c}(t)$, is represented mathematically after the initial linear phase of duration t_0 elapses, then we have

$$\bar{c}(t) = ae^{-bt} \tag{6.84}$$

where a and b are dimensional constants. Comparing Eqs. 6.83 and 6.84, we obtain the following relations

$$a = \frac{m}{V}e^{\frac{k}{V}t_0}, \quad b = \frac{k}{V}. \tag{6.85}$$

Calculating the area A under the experimental curve and comparing it with the theoretical value, we obtain

$$A = \int_0^\infty \bar{c}(t)\,dt$$

$$= \int_0^\infty c(t)\,dt$$

$$= \frac{m}{V} \int_{t_0}^\infty e^{-\frac{k}{V}(t-t_0)}\,dt$$

$$= \frac{m}{k}, \quad t > t_0. \tag{6.86}$$

Thus, we obtain

$$k = \frac{m}{A} = \frac{m}{\int_0^\infty \bar{c}(t)\,dt},$$

$$V = \frac{m}{Ab},$$

$$t_0 = \frac{1}{b}\log\left(\frac{a}{Ab}\right). \tag{6.87}$$

The theoretical curve thus obtained with the value of t_0 obtained from above is shown in Fig. 6.7.

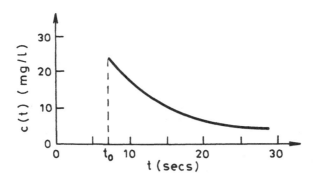

Fig. 6.7: The theoretical curve of dye concentration.

However, the typical dye concentration curve recorded from a human subject looks like Fig. 6.8. Because of recirculation of dye, the concentration does not fall-off as in the theoretical curve. The dashed line on the descending curve represents the semilogarithmic extrapolation of the upper portion of the descending part, prior to the beginning of recirculation.

From Eq. 6.87, we conclude that the cardiac output k is given by

$$k = \frac{\text{mass of indicator injected}}{\text{area under the indicator-dilution curve}}. \tag{6.88}$$

This result is usually referred to as *Hamilton's formula*.

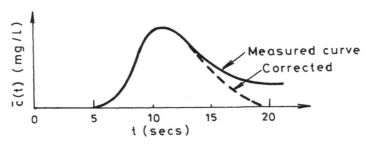

Fig. 6.8: A typical indicator dilution curve.

6.5 Peristaltic Flows

6.5.1 *Introduction*

The word peristalsis stems from the Greek word *peristaltikos*, which means clasping and compressing. Peristalsis is a muscle-controlled flow similar to the flow in the cardiovascular system. Peristalsis occurs in many organs. Peristaltic flow is the flow generated in the fluid contained in a distensible tube when a progressive wave travels along the wall of the tube. Although the elasticity of the wall does not directly enter into the flow equations, it affects the flow through the progressive wave travelling along its length. Peristaltic flow occurs widely in the functioning of ureter, transport of spermatozoa in the cervical canal, transport of bile, transport of cilia, vaso motion of small blood vessels, etc. Of these physiological flows, peristalsis in ureter, intestine and stomach are the most important, from the biofluid mechanics point of view. The mathematical problem of peristaltic flow is similar to that of the collapsible tube. In the case of functioning of the ureter, we model it mathematically assuming that the ureter receives fluid from the kidney at the upper end, and passes it down to the bladder against a pressure gradient. Normally, there is more than one wave along the entire length of the ureter, which is of the order of 30 cm. The amplitude of the wave is of the order of 5 mm and its speed is about 6 cm/sec. The frequency of contractions is about 1 to 8 per minute. Each contraction lasts about 1.5 to 9 sec, the diastolic (expansion) phase is about twice as long as the systolic (contraction) phase. Pressure during the contraction varies from 2 to 8 mm Hg at the pelvis, 2 to 10 mm Hg in the upper portion of the ureter, and 2 to 14 mm Hg in the lower portion (see Weinberg [70]).

6.5.2 *Peristaltic Motion in a Cylindrical Tube*

On the basis of peristaltic transport in the ureter described in Section 6.5.1, let us now consider an idealized two-dimensional peristaltic problem of studying the wave motion of the wall of a circular cylindrical tube at a moderate amplitude. A sinusoidal wave is assumed to travel down the walls of a two-dimensional channel of constant width and infinite length. We assume the cylindrical tube or the channel is filled with a homogeneous Newtonian viscous fluid, although the fluid involved may be non-Newtonian, and the flow may take place in two layers, ie., a core layer and a peripheral layer. Taking the x-axis along the center line of the channel, and the y-axis normal to it, the equations governing two-dimensional motion of a viscous fluid, the continuity and Navier-Stokes equations, are given by (see Section 1.3)

$$\frac{\partial u}{\partial x} + \frac{\partial v}{\partial y} = 0, \tag{6.89}$$

$$\frac{\partial u}{\partial t} + u\frac{\partial u}{\partial x} + v\frac{\partial u}{\partial y} = -\frac{1}{\rho}\frac{\partial p}{\partial x} + \nu\nabla^2 u, \tag{6.90}$$

$$\frac{\partial v}{\partial t} + u\frac{\partial v}{\partial x} + v\frac{\partial v}{\partial y} = -\frac{1}{\rho}\frac{\partial p}{\partial y} + \nu\nabla^2 v. \tag{6.91}$$

If we use the stream function ψ as given in Eq. 1.36, ie., $u = \psi_y$, $v = -\psi_x$, and eliminate the pressure terms in Eqs. 6.90 and 6.91 we obtain as in Eq. 1.38

$$\nabla^2\psi_t + \psi_y\nabla^2\psi_x - \psi_x\nabla^2\psi_y = \nu\nabla^4\psi \tag{6.92}$$

in which

$$\nabla^2 \equiv \frac{\partial^2}{\partial x^2} + \frac{\partial^2}{\partial y^2}. \tag{6.93}$$

We assume the fluid is subjected to conditions imposed by the symmetric motion of the elastic walls. Let the vertical displacements of the upper and lower walls be η and $-\eta$, respectively. Hence, the equations of the walls are given by

$$y = \pm\eta(x,t) = \pm a\left[1 + \epsilon\cos\frac{2\pi}{\lambda}(x - ct)\right] \tag{6.94}$$

where ϵ is the amplitude ratio, λ is the wavelength, c is the wave speed and a is the undeformed radius of the tube (see Fig. 6.9).

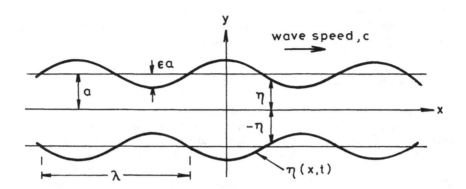

Fig. 6.9: Peristaltic motion of the flexible walls.

The main objectives in this study are to determine the longitudinal pressure gradient that can be generated by the travelling wave, and the flow resulting from peristalsis superimposed on pressure differences at the ends of the tube.

In order to solve Eq. 6.92 we assume the following boundary conditions

$$u = 0$$

and

$$v = \pm \frac{2\pi a c \epsilon}{\lambda} \sin \frac{2\pi}{\lambda}(x - ct) \quad \text{at} \quad y = \pm \eta(x, t). \tag{6.95}$$

Introducing the dimensionless quantities

$$X = \frac{x}{\lambda}, \quad Y = \frac{y}{a}, \quad T = \frac{ct}{\lambda}, \quad \Psi = \frac{\psi}{ac},$$

$$n = \frac{a}{\lambda}, \quad R_e = \frac{ac}{\nu}, \tag{6.96}$$

Eq. 6.92 finally reduces to

$$D_n^2 \Psi_T + \Psi_Y D_n^2 \Psi_X - \Psi_X D_n^2 \Psi_Y = \frac{1}{nR_e} D_n^2 \Psi \tag{6.97}$$

where the operator D_n is given by

$$D_n^2 \equiv n^2 \frac{\partial^2}{\partial X^2} + \frac{\partial^2}{\partial Y^2}. \tag{6.98}$$

The corresponding boundary conditions become

$$\Psi_Y = 0,$$

and

$$\Psi_X = 2\pi\epsilon \sin 2\pi(X - T) \quad \text{at} \quad Y = \pm\eta(X,T). \tag{6.99}$$

From the above, dimensionless quantities, we notice the following:

(i) The Reynolds number R_e will be small if the wave speed is small or the distance between the walls is small or the kinematic viscosity is large.

(ii) The wave number n will be small if the wavelength is large as compared to the distance between the walls.

(iii) The amplitude ratio ϵ will be small if the amplitude of travelling waves (displacement of the wall) is small as compared to the distance between the walls.

In order to solve the problem, various approximations are made such as

(i) small Reynolds number so that nonlinear convective terms in the Navier-Stokes equation can be neglected,

(ii) long-wavelength compared with the undeformed radius of the tube,

(iii) small amplitude of the wall displacement compared with the tube radius.

6.5.3 *Long-Wavelength Analysis*

Let us make the assumptions of small Reynolds number and small wave number, ie.,

$$R_e \ll 1, \quad n \ll 1. \tag{6.100}$$

Consider cylindrical coordinates such that the Z-axis is along the centre-line of the tube. The equation of the tube wall is given by

$$h(Z,t) = a\left(1 + \epsilon \sin \frac{2\pi}{\lambda}(Z - ct)\right) \tag{6.101}$$

where, as before, a is the undeformed radius of the tube (see Fig. 6.10). We assume that the pressure is independent of the radial coordinate, ie.,

$$p = p(Z,t). \tag{6.102}$$

We introduce a moving coordinate system (r, z) travelling with the wave (see Kapur [33]) so that

$$r = R, \quad z = Z - ct. \tag{6.103}$$

We have the continuity equation and equation of motion as given before in Eqs. 1.44 and 1.46

$$\frac{\partial}{\partial r}(ru) + \frac{\partial}{\partial z}(rw) = 0, \tag{6.104}$$

$$\frac{\partial p}{\partial z} = \mu \left(\frac{\partial^2 w}{\partial r^2} + \frac{1}{r} \frac{\partial w}{\partial r} \right) \tag{6.105}$$

where u and w are the radial and axial velocity components of the fluid in the moving coordinate system.

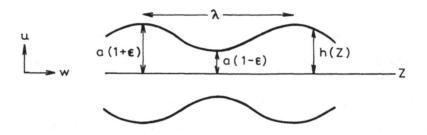

Fig. 6.10: The geometry of the tube.

Using the boundary condition

$$w = -c \quad \text{at} \quad r = h \tag{6.106}$$

we get from Eq. 6.105

$$w = -c - \frac{1}{4\mu} \frac{\partial p}{\partial z}(h^2 - r^2). \tag{6.107}$$

The flow rate is given by

$$q = 2\pi \int_0^h rw \, dr$$

$$= -\pi h^2 c - \frac{\pi h^4}{8\mu} \frac{\partial p}{\partial z}. \tag{6.108}$$

So, we have

$$\frac{\partial p}{\partial z} = -\frac{8\mu q}{\pi h^4} - \frac{8\mu c}{h^2}. \tag{6.109}$$

We thus obtain from Eq. 6.107

$$w = -c + 2\left(\frac{q}{\pi h^4} + \frac{c}{h^2}\right)(h^2 - r^2). \tag{6.110}$$

For radial velocity u, we integrate Eq. 6.104 and obtain

$$ru = -\int_0^r r\frac{\partial w}{\partial z}\,dr, \tag{6.111}$$

which with the help of Eq. 6.110 reduces to

$$u = -\frac{\partial h}{\partial z}\left(\frac{cr^3}{h^3} - \frac{2qr}{\pi h^3} + \frac{2qr^3}{\pi h^5}\right). \tag{6.112}$$

Reverting back to the stationary system, the velocity components U and W, and the flow rate Q are given by

$$U = u = -\frac{\partial h}{\partial Z}\left(\frac{cR^3}{h^3} - \frac{2qR}{\pi h^3} + \frac{2qR^3}{\pi h^5}\right)$$

$$= -\frac{2\pi a\epsilon}{\lambda}\cos\frac{2\pi}{\lambda}(Z - ct)\left(\frac{cR^3}{h^3} - \frac{2qR}{\pi h^3} + \frac{2qR^3}{\pi h^5}\right) \tag{6.113}$$

$$W = w + c = 2\left(\frac{q}{\pi h^4} + \frac{c}{h^2}\right)(h^2 - R^2) \tag{6.114}$$

$$Q = 2\pi\int_0^h WR\,dR = q + \pi ch^2 \tag{6.115}$$

where h is a function of Z and t given by Eq. 6.101.

To determine the pressure drop across one wavelength λ, we have from Eq. 6.109, after integrating,

$$(\nabla p)_\lambda = -\frac{8\mu q}{\pi a^4}\int_0^\lambda \frac{dz}{\left(1 + \epsilon\sin\frac{2\pi}{\lambda}z\right)^4} - \frac{8\mu c}{a^2}\int_0^\lambda \frac{dz}{\left(1 + \epsilon\sin\frac{2\pi}{\lambda}z\right)^2}$$

$$= -\frac{4\mu}{\pi a^4}\left[\frac{q(2 + 3\epsilon^2)}{(1 - \epsilon^2)^{\frac{7}{2}}} + \frac{2\pi ca^2}{(1 - \epsilon^2)^{\frac{3}{2}}}\right]. \tag{6.116}$$

Clearly, the pressure drop across one wavelength would be zero if

$$q = -\frac{2\pi ca^2(1 - \epsilon^2)^2}{(2 + 3\epsilon^2)} \tag{6.117}$$

which gives the corresponding velocity components

$$U = -\frac{2\pi a\epsilon cR}{\lambda h^3}\left[R^2 + \frac{4ca^2(1 - \epsilon^2)^2}{2 + 3\epsilon^2}\left(1 - \frac{R^2}{h^2}\right)\right]\cos\frac{2\pi}{\lambda}(Z - ct), \tag{6.118}$$

$$W \;=\; 2c \left(1 - \frac{2a^2(1-\epsilon^2)^2}{h^2(2+3\epsilon^2)}\right)\left(1 - \frac{R^2}{h^2}\right). \tag{6.119}$$

Clearly, when $\epsilon = 0$, the results are that of a cylindrical tube of uniform cross-section.

Chapter 7

FLUID MECHANICS OF HEART VALVES

7.1 General Introduction

The heart valves are mechanical devices that permit the flow of blood in one direction only. The two cuspid (atrio-ventricular) valves are located between the atria and ventricles, while the two semi-lunar valves are located at the entrance to the pulmonary artery and the great aorta. The cuspid valves prevent blood from flowing back into the atria from the ventricles and the semi-lunar valves prevent it from regurgitating into the ventricles from the aorta and pulmonary artery.

Any one of the four valves may lose its ability to close tightly; such a condition is known as valvular incompetency or regurgitation. On the other hand, stenosis is an abnormality in which the fully-open valvular orifice becomes narrowed by scar tissue that forms as a result of valvular disease; thereby the blood flow through the valve results in increased pressure drop across the valve orifice.

Heart valve disease in advanced forms severely affects the performance of a natural heart valve by causing permanent alteration to the valve tissue leading to valve dysfunction. An improper functioning heart valve induces considerable disability and may ultimately lead to death if the problem is not treated. The need for surgical repair or total valvular replacement arises when a patient's quality of life suffers from the ensuing disability. Biological and prosthetic valve substitutes have been employed since the early 1960s when valve replacement was first introduced as a clinical procedure. Today, mechanical prostheses and tissue valves (bioprostheses) are in widespread clinical use in medical centres throughout the world.

129

Despite over 25 years of development and clinical experience prosthetic heart valves continue to suffer from numerous pathological problems. The ideal valve substitute has yet to be designed. Although many difficulties can be expected during and immediately following a valve implant, the most debilitating and life-threatening complications are those that begin to develop from the moment of implant and manifest themselves in latter years.

The presence of the prosthesis disturbs the flow of blood producing areas of high fluid stress, high wall stress and separated flow regions. Depending upon their extent, these flow conditions may eventually cause many serious pathological problems including sub-haemolytic injury and haemolysis, endothelial damage and erosion, thrombotic deposition and related thromboembolic events. There is a direct link between the fluid dynamics of heart valves and the pathological problems encountered. Hence, it is extremely important to examine the likelihood of these problems by investigating as fully as possible the fluid dynamics of prosthetic heart valves.

Measurements using *in vitro* experiments provide much valuable information. However, physical measurements are often limited in application due to probe size or are simply not suitable for many regions of the flow domain. This has been the experience of previous researchers in this field (Figiola and Mueller [15]; Herkes and Lloyd [26]). Experimental measurements are unable to provide a complete picture of the complex flow patterns.

Computational fluid dynamic techniques, that will be discussed in the next chapter, provide a powerful noninvasive tool for obtaining good approximations to velocity and stress distributions in a flow field. However, depending upon the nature of the fluid dynamics, the mathematical equations and the discrete modelling, the numerical solutions can range from very accurate to only rough approximations. Current numerical techniques provide acceptably good results for most physical flow problems. Only the more complex flow cases (eg., turbulent pulsatile flow) cause difficulties. The advent of high speed digital computers and ready access to modern computing power has generated strong interest in the development of numerical models to investigate prosthetic valve fluid dynamics. A number of important studies in this field have for the most part concentrated on laminar flow and/or idealized valve geometries. However, in reality, a complex combination of both laminar and turbulent flow and its relaminarization is observed through prosthetic valves during the cardiac cycle. Furthermore, the tissue walls are generally curved.

In the next chapter, the application of computational methods to elucidate the complex fluid dynamics associated with steady flow through prosthetic heart valves, will be examined by including turbulence effects and accurate boundary shape.

7.2 A Brief Description of the Heart Valves

As we know, the human heart is a pulsatile synchronous pump which beats more than 40 million times each year. Its function is to provide the mechanism by which the oxygen requirements of the human body are satisfied by maintaining the circulation of blood.

It has been mentioned in Chapter 2, that the human heart has four chambers — two chambers pumping blood and two chambers receiving blood. The right and the left ventricles are the two separate pumping chambers of the heart, pumping blood to the lungs and the body through two large blood vessels. The vessel supplying the lungs is the pulmonary artery, and that supplying the body is the aorta. The two receiving chambers in the heart are the left and right atria. The one receiving oxygenated blood from the lungs is the left atrium, and that receiving deoxygenated blood from the veins of the body is the right atrium.

The human heart has four valves, two on each side. The valves on the right side are the tricuspid valve — an inlet valve, and the pulmonary valve — an outlet valve. The valves on the left side are the mitral valve — an inlet valve, and the aortic valve — an outlet valve. In the natural heart, the valve leaflets open at the centre to allow unobstructed central flow. The action of the four heart valves within the heart acts to restrict blood flow to one direction and prevent substantial backflow.

The aortic and pulmonary valves each consist of three cusps attached to a fibrous tissue ring on the inner walls of the aorta and pulmonary artery, respectively. Both outlet valves have similar dimensions. The adult aortic or pulmonary valve is approximately 25 mm in diameter. Behind each cusp is a pouch-like cavity embedded into the artery wall. These regions are known as the sinuses of Valsalva. An important mechanism in aortic valve closure is the production of vortices in the region between each cusp and its sinus. Bellhouse and Talbot [5] demonstrated this behaviour by constructing a trileaflet valve mounted in a rigid chamber with a model aortic sinus.

The mitral and tricuspid valves are also similar to each other. However, the mitral valve has two major and two minor leaflets, whereas the tricuspid valve has three similarly sized leaflets. Unlike the other valves, the circumference of the mitral valve is not circular but "D-shaped", measuring approximately 35 mm across the major diameter. A cutaway view of the heart illustrating anatomical relationships is shown in Fig. 7.1.

After circulating through the body, deoxygenated blood collects in the great veins. It then flows into the right atrium and through the tricuspid valve into the right ventricle. The right ventricle contracts, and together with the closure of the tricuspid valve, the deoxygenated blood is ejected through the pulmonary valve into the pulmonary artery. The blood flows to the lungs where the transfer of carbon dioxide and oxygen occurs. Four pulmonary veins then direct the reoxygenated blood into the left atrium where the blood flows through the mitral valve into the left ventricle. This filling phase of the cardiac cycle is known as diastole.

When the diastolic cycle is complete, the left ventricle contracts abruptly, the mitral valve closes, and blood is ejected through the aortic valve. This phase of the cardiac cycle is known as ventricular systole. Left and right ventricular systole occur simultaneously as do left and right diastole. The muscular contractions are controlled by a group of cells situated in the heart. The blood which is ejected through the aortic valve passes down the aorta into the arteries and then to the veins which reach most of the body. A schematic diagram of blood flow within the heart is shown in Fig. 7.2 (see also Fig. 2.3, page 21).

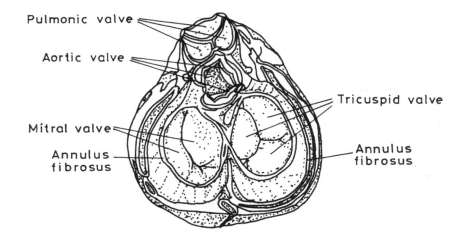

Fig. 7.1: Drawing of the four cardiac valves as viewed from the base of the heart. Note the manner in which the leaflets overlap in the closed valves.

The left and right coronary arteries supply oxygenated blood to the muscle of the heart known as the myocardium. These arteries originate from two of the three sinuses of Valsalva located at the base of the aorta. The flow of blood through these arteries is crucial for the proper functioning of the heart muscle.

The volume of blood passing down the aorta during each systole equals the left ventricular stroke volume, minus the amount which flows into the coronary arteries, minus the amount of retrograde flow through the aortic and mitral valves. Cardiac output in man varies between about 5 litres min^{-1} at rest to about 25 litres min^{-1} during extreme exertion. Normally retrograde flow is negligible except in diseased valves where it may be considerable. The stroke volume may then become several times greater than the flow down the aorta. This is clearly undesirable as the cardiac output of a healthy individual is normally sufficient to fulfill the oxygen requirements of the body.

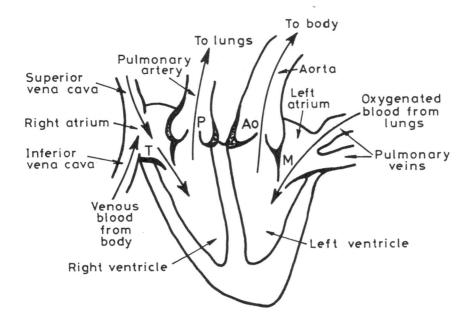

Fig. 7.2: Schematic diagram of the blood flow through the heart valves. The arrows show the direction of blood flow. The valves are indicated by symbols: T, tricuspid; P, pulmonary; Ao, aortic; and M, mitral.

A valve is called incompetent when it fails to close properly, causing a jet of blood to flow retrogradely through the valve when it should be shut. Both mitral and aortic incompetence are common valvular defects.

Another common valve defect is stenosis. This is a narrowing of the valve which restricts the flow of blood across the valve. Increased pressure drops are consequently induced across a stenosed valve. In mitral stenosis this pressure increase is transmitted back through the lungs to the pulmonary artery. The body may then try to compensate for these effects by reducing cardiac output making physical exertion difficult and uncomfortable. Irreversible damage to lung function and right ventricular failure have been known to occur as a result of prolonged severe mitral stenosis. Aortic stenosis also occurs causing a raised left ventricular systolic and a reduced aortic pressure. The effect of blood flowing through a stenosed aortic valve at high pressure is to cause turbulence in the aorta, aortic dilation and deposition of calcium on the aortic wall and on the valve leaflets. Tricuspid stenosis and incompetence often occurs in association with mitral valve stenosis and incompetence, respectively.

The human heart valves may become defective through acquired disease or because of congenital malformation. Congenital defects may initially appear minor, but after many years they can become quite serious. The most common acquired diseases which may cause valvular defects are rheumatic valvular heart disease and bacterial endocarditis. Even though the incidence of rheumatic heart disease is slowly decreasing, it remains present in over 1.5 million adults in the United States, and is responsible for about 15,000 deaths each year in subjects under 65 years of age. The major damage to the heart from this disease results from gradual scarring of the heart valves which may take many years to become evident. A bacterial invasion of the blood stream by a simple wound could cause gross damage to the heart valve leaflets if circulating bacteria lodge on a small valve lesion. Both of these diseases can result in valvular incompetence, while stenosis can also result from rheumatic valve disease.

Papillary muscle dysfunction can occur due to blockage of coronary artery branches to the papillary muscles of the mitral valve, resulting in sudden mitral incompetence. Even a myocardial infarct can sometimes injure the attachments of the valves to the wall.

It is clear that if incompetence or stenosis is present in heart valves, then serious consequences can ensue for the heart and circulation. The need for valvular replacement or surgical repair arises when these conditions threaten the life and well-being of the patient.

7.3 Prosthetic Heart Valves

The preservation of proper circulationn is of paramount importance to patient well being, and this often necessitates the replacement of a dysfunctioning valve, or valves, by artificial substitutes. Today, nearly 75000 prosthetic heart valves are implanted annually throughout the world.

Let us examine many important aspects of prosthetic heart valves. In particular, the history of valve replacement, a general description of prosthetic valves, and the numerous pathological complications that may and very often do arise following a valve implant.

7.3.1 *History of Valve Replacement*

Prior to the development of prosthetic valves, the surgical correction of valvular defects carried a high mortality rate and any benefits obtained were usually only short-lived. With mitral stenosis, surgical treatment was directed towards conversion of the stenosis into incompetence. The first successful treatment was performed by Souttar [59] in 1925, who used his fingers to dilate a stenotic mitral valve. Many years later the second successful treatment of this type was performed by Bailey [3] in 1949. By the early 1950s, closed valvulotomy, which consisted of blindly inserting the finger through the atrium or ventricle and relieving the stenotic valve in the beating heart, had become a safe and successful procedure. Dilatation of stenotic valves using fingers was almost entirely discontinued in 1959 with the introduction of the transventricular dilator. This mechanical dilator was appreciated for providing much less general disturbance to the heart.

A new era in cardiac surgery emerged with the development of the heart-lung machine for the extra-corporeal circulation of the blood. It had become possible to operate on the completely arrested open heart while the heart-lung machine served to supply oxygenated blood to the rest of the body. As a direct consequence of this development, a great number of new possibilities became available in valvular repair. The pathway for total valve replacement had been opened and the design of suitable valve replacements constituted a major cardiac research project during the 1960s. The primary consideration of the valve designers involved in the initial development was prevention of any serious regurgitation in the closed valve.

In 1953 the first prosthetic heart valve was implanted by Hufnagel without the use of a heart-lung machine (Hufnagel *et. al.* [27]). The actual insertion time required only between 3 and 6 minutes. The valve was of the caged-ball type and was placed into the descending aorta of patients for the treatment of severe aortic incompetence. Locating the valve in a more physiologic position was simply too difficult. Unfortunately, the valve was unable to prevent regurgitation from the aortic arch vessels and was also prone to a high incidence of thromboembolic events.

In 1960, Harken *et. al.* [24] performed the first subcoronary insertion of a caged-ball valve in a group of patients with aortic regurgitation. Although the caged-ball prosthesis was at first employed unsuccessfully in the aortic position it was successfully used for mitral valve replacement by Starr and Edwards [60] in the same year. Following a great number of modifications, the caged-ball prosthesis continues to remain in use today. However, its centrally obstructive design gives it only adequate haemodynamics and is responsible for thrombogenic events and mild haemolysis.

The caged-disk valve was introduced in 1965 by several investigators in an attempt to improve haemodynamics and provide a lower profile and lighter weight then the caged-ball valve (Hufnagel and Conrad [28]; Kay *et. al.* [35]). Unfortunately, the caged-disk design is the most flow obstructive of all prosthetic valves, particularly in the aortic position where it has been shown to increase pre-existing mitral insufficiency. Other problems include high thrombogenicity, haemolysis, intimal proliferation in the aortic root and disk cocking.

In an attempt to overcome these difficulties the tilting-disk prosthesis was introduced in 1967 by Wada. Lillehei-Kaster and Björk-Shiley tilting-disk valves were introduced soon after (Kaster-Lillehei [34]; Björk [7]). Initial problems with disk wear caused by the hinge struts in the Björk-Shiley prosthesis were overcome by incorporating into the design the ability for the disk to turn freely once every 180 to 200 cycles. The potential for disk swelling was removed with the introduction of the pyrolytic carbon disk in 1972. A recent series of the tilting-disk valve use a convex-concave disk which allows greater forward flow by rising further out of its housing when open. The semi-central nature of flow through tilting-disk valves and excellent haemodynamics have made this valve a popular choice with cardiac surgeons [52].

St. Jude Medical, Inc. introduced a bileaflet pivoting-disk prosthesis for clinical trial in 1977. The valve consists of two hinged semi-circular leaflets housed in an orifice ring which have an opening angle of 85° and a closing angle of 30° to 35°. The pivoting mechanism eliminates the need for the supporting struts designed into both the Björk-Shiley and Lillehei-Kaster tilting-disk valves. When open the St. Jude valve provides a superior high-flow orifice with nearly complete centralised flow. The St. Jude prosthesis moved from clinical investigation status to clinical use in early 1983 with a complication rate as low or lower than that of any other mechanical valve available.

Several cardiac surgeons have preferred to use biological substitutes for valve replacement. Homograft valves (removed from a cadaver), heterograft valves (removed from a dead animal), or tissue-fabricated valves (constructed from nonvalvular human and animal tissue) have been used extensively for the replacement of diseased aortic and mitral valves. Tissue-fabricated valves are designed to duplicate the flow behaviour of natural valves. Although short-term results are good, the long-term results for tissue bioprosthetic valves are not satisfactory due to calcification, leaflet degeneration and eventual valve stenosis. The main advantage of employing tissue valves over synthetic mechanical valves is their almost complete freedom from thrombus formation. To this date, no mechanical or biological valve replacement has been developed which is completely satisfactory.

Nearly 50 different prosthetic heart valve designs have been developed since the early 1960s. The vast majority of these valves have been discarded and, of those remaining, several modifications have been made to improve haemodynamic performance, wear characteristics and overall valve function. At present there are five major categories of prosthetic valves in use: (1) Caged-ball; (2) caged-disk; (3) tilting-disk; (4) bileaflet pivoting-disk; and (5) tissue bioprosthesis.

A typical mechanical prosthesis consists of four basic components: (1) Occluder; (2) cage, strut or hinge; (3) sewing ring; and (4) orifice ring. The occluder responds passively to pressure and flow changes within the heart to allow forward flow and to prevent substantial backflow. It is free to move under action of blood flow. The cage, strut or hinge retains and guides the motion of the occluder. The prosthesis is attached into the heart by sutures which pass through the sewing ring, into the tissues and back through the sewing ring. Both the cage (or struts) and sewing ring are attached to the orifice ring. Blood flows through the orifice when the valve is open and is sealed by the occluder when the valve closes (see·Fig. 7.3).

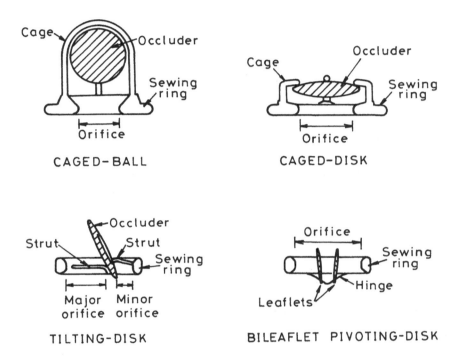

Fig. 7.3: Diagrammatic representation of major categories of mechanical prosthetic heart valves in use today. The valves are fully open and key components are identified.

There are numerous prosthesis-related complications that can arise following a valve implant. Fatal and non-fatal problems which are unrelated to valve fluid dynamics include infection (prosthetic valve endocarditis), haemorrhage during anticoagulation therapy, mechanical malfunction due to material degradation, separation of prosthesis from its attachment site (valve dehiscence) and interference with cardiac function.

Other non-fatal conditions related to valve fluid dynamics which may cause concern include haemolysis, sub-haemolytic injury, endothelial damage and tissue overgrowth. It is clear, therefore, that the fluid dynamics of prosthetic valves play a crucial role in the aetiology of several important prosthesis-related complications.

Let us now discuss a few prosthesis-related complications attributable to valve fluid dynamics.

7.3.2 *Thrombosis and Thromboembolism*

Thrombosis is defined as the formation of a blood coagulum within a vessel or the heart. There are several factors which favour thrombus formation. These include:

(1) Stasis, or reduction in velocity of blood flow, which is common in varicose veins and inactivity;

(2) derangement in composition of blood as may be experienced through disease, surgery or childbirth;

(3) recirculating, or separated flow caused by disruption to normal flow by the presence of a flow constriction as in a heart valve prosthesis; and

(4) alterations in the endothelium which may be due to inflammation, vascular diseases or elevated wall shear stresses.

Such flow conditions prevail at various locations in the vicinity of heart valve prostheses. During systole, a predilection site for thrombotic deposition occurs near the flow reattachment point of the sinus recirculation zone and along the downstream face of the sewing ring. In the large circulation zone, the downstream side of an occluder-type valve (eg., caged-disk or caged-ball) is also a region of potential thrombus formation. The presence of a cage or strut further enhances the possibility of thrombogenesis. Varying amounts of thrombus formation have been observed on recovered Starr-Edwards aortic caged-ball valves with a predominance at the base of the cage struts and the apex of the cage. A schematic example of thrombus formation and tissue overgrowth is shown in Fig. 7.4.

The complications that can arise from thrombosis of a prosthetic valve are:

(1) Thrombotic stenosis, when the prosthesis is partially blocked with a thrombus thus inhibiting the valve's performance; and

(2) thromboembolic events, when part of the thrombus, known as an embolus, breaks off the main thrombus body and is carried by the blood to lodge in the small arteries of a vital organ such as the heart, kidney or brain causing temporary or permanent damage.

The incidence of thromboembolism from prosthetic valves represents a major threat to patients. Consequently, intense anticoagulation therapy is mandatory with any mechanical valve implant.

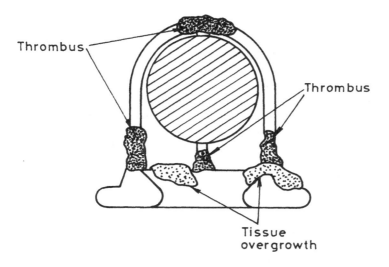

Thrombus

Thrombus

Tissue
overgrowth

Fig. 7.4: Diagrammatic example of thrombus formation and tissue overgrowth on a Starr-Edwards caged-ball valve.

7.3.3 *Haemolysis*

The basic characteristics of shear flow apply not only to pure fluids but also to complex suspensions like blood. Locally accelerated blood flow occurs readily in the vicinity of prosthetic heart valves inducing abnormally high fluid forces, or shear stress. A blood cell which is subjected to a region of shearing action exerted by the surrounding fluid will experience a distribution of shear stress over its entire membrane. Consequently, the cell membrane will be stretched and may suffer irreversible changes detrimental to its essential function.

Sub-haemolytic injury, or damage to blood cells, is common in prosthetic valve recipients but its effects can usually be compensated by the body's ability to substantially increase its rate of haemoglobin production. If the rate of destruction exceeds the rate of production, clinically significant anaemia will be observed. The commonest cause of haemolytic anaemia is the development of a leak between the sewing ring and its attachment site. This leak is physically similar to jet-like flow and induces extremely high shearing forces on the blood. The prosthesis itself continues to damage blood cells by its mere presence in the flow field, with significant anaemia usually absent.

Estimates for shear induced in-bulk haemolysis thresholds have varied widely, from 150 to over 4000 Nm^{-2}. Several reasons for these apparent discrepancies have been proposed including the extent to which containing surfaces are responsible for the observed levels of haemolysis. Leverett *et. al.* [38] proposed that shear stress and exposure time are the two primary determinants for in-bulk haemolysis. Cells exposed for very short durations to very high stresses do not lyse, but exposure for long durations at much lower stress levels can cause lysis. However, below a certain threshold of fluid stress normal red blood cells will not deform and rupture even if exposed for an indefinite period. Two independent studies, one performed with turbulent shear flow (Sutera and Mehrjardi [64]) and the other with laminar shear flow (Leverett *et. al.* [38]) have shown this low stress threshold is between 150 and 250 Nm^{-2}. It can also be inferred that the presence of the fluctuating characteristics of turbulent flow are not harmful provided the time-averaged shear stress does not exceed the threshold.

7.3.4 *Endothelial Damage*

The endothelium is a delicate membrane which lines the heart, blood vessels and body cavities. In large arteries such as the aorta, rapidly flowing blood exerts a force, or viscous drag, upon the endothelial lining of the artery which attempts to pull the lining along with the flow. This force, or shear stress, is directly proportional to the velocity gradient near the vessel wall. Normally the levels of wall shear stress are so small that no damage or erosion of endothelial cells occurs. However, during systole, the peripheral flow induced by caged-disk and caged-ball valves, and the semi-central flow through tilting-disk prostheses causes a significant increase in velocity gradients at the proximal portions of the aorta. As has already been mentioned, damage to the endothelium also acts to increase the potential for thrombus formation by increasing the "stickiness" of the vessel wall. It is therefore clear that a highly desirable feature of any valve prosthesis should be low levels of wall shear stress.

7.3.5 *Tissue Overgrowth*

Tissue overgrowth is often seen on valve prostheses recovered during autopsy. Yoganathan *et. al.* [73] observed tissue overgrowth on the aortic side of the sewing ring and on the struts of cloth-covered Starr-Edwards caged-ball valves (see Fig.

7.4). The cloth covering (eg., polypropylene mesh) on cage struts and valve orifice has reduced the incidence of thromboembolism by allowing infiltration of fibrous tissue which eventually serves as a nonthrombogenic surface. However, excessive tissue growth into the cloth covering will eventually result in valve stenosis and an increase in the pressure drop across the valve, thus diminishing hydraulic performance. The development of less thrombogenic materials (eg., pyrolytic carbon) for use in valve construction has led to some progress in the control of tissue overgrowth. Nevertheless, tissue overgrowth is not limited to cloth covered valves where it has been primarily noticed, but appears to depend to some extent on the fluid mechanics of the valve, although this process is not fully understood. High blood flow and the associated effects of fluid scouring may play a role in limiting or preventing tissue overgrowth.

Chapter 8

COMPUTATIONAL BIOFLUID MECHANICS

8.1 Introduction

The development of computational methods to solve nonlinear partial differential equations has been progressing for over 60 years. The advent of high speed digital computers has seen many advances in this field by enabling development, comparison and refinement of the numerical techniques. A historical outline of computational fluid dynamics is available in Roache [54]. Publications of Journals such as *Journal of Computational Physics, Computers and Fluids, Journal of Fluid Mechanics* and *Numerical Methods in Engineering* contain papers describing the application of numerical methods to numerous physical problems including physiological flow processes.

One particular method, the *finite difference method*, remains the predominant technique for obtaining numerical solutions to fluid flow problems. The method can be summarised by the following stages (Noye [48]):

(1) A grid is developed to cover the solution region. At the grid points (ie., intersection of grid lines) the approximate solution to the problem will be found.

(2) All dependent variables are initialised. The initial values are either known exactly, or they may at best be only a rough guess, especially if they require determination through application of the solution procedure.

(3) The finite difference equations corresponding to the continuous partial differential equations are constructed. They are used to calculate approximate values of the dependent variables at all grid points in the solution domain.

(4) The computational cycle commences. New values are calculated at adjoining space positions by incrementing some value Δx, and at subsequent time levels by Δt. The process is continued until a predetermined time is reached, or as in the steady state, the dependent variables at each grid point remain unchanged at successive time levels.

Unfortunately, it is generally impossible to obtain theoretical estimates of the accuracy, convergence and stability of a numerical solution to a problem possessing large nonlinear terms in the governing equations. This is especially true for viscous fluid flow problems with large Reynolds numbers. In order to verify the accuracy of finite difference techniques applied to such problems, it is necessary to compare numerical solutions with experimental results and/or employ computational experimentation such as observing solution behaviour when subject to grid refinement. The mathematical theory for convergence and stability of numerical schemes have only been developed for linear systems. Consequently, results from linear theory are used to guide nonlinear problems but many *ad hoc* approaches remain a practical necessity. For example, with fluid flow problems, it is often necessary to anticipate the nature of the solution in advance so that a larger number of grid points can be assigned to regions of interest, particularly those with high velocity gradients. Typically, these regions occur at constrictions and near wall boundaries. Also, specification of iterative relaxation parameters normally requires some testing beforehand to determine optimal values. It may be found that switching to slightly different values after a certain number of global iterations serves to enhance convergence and control stability. Alternating the direction of iterative sweeps also seems to provide additional solution control. Monitoring solution progress is mandatory and its implementation depends upon the nature of the problem, the availability of experimental results for comparison and the degree of interactiveness of the computing environment. In this regard, computer graphics are very useful.

The vast majority of problems associated with computational fluid dynamics are solved using finite difference methods and a regular Cartesian coordinate system. Consequently, flow domain geometry is often simplified by making boundary walls coincide with the rectangular computational mesh. In a sense, the physical domain is fitted to the computing mesh by adjusting the location of grid lines until the flow domain is satisfactorily defined. Clearly, this approach is not suited for application to more general problems, particularly those where curved physical boundaries strongly influence the flow characteristics. Domains with greater geometrical complexity are often solved with the more formidable finite element method and demand great programming effort when compared to the finite difference approach. A technique that embodies the computational efficiency of finite differences and facilitates application to general body problems is the numerical generation of *boundary-fitted coordinates*.

In a boundary-fitted coordinate system, the physical domain (z,r) is numerically transformed, or mapped, onto a calculation domain in coordinated (ξ, η) — a regular rectangular coordinate system (Fig. 8.1). All numerical calculations involving the governing equations are conveniently performed in the transformed (ξ, η) plane rather than the physical (z,r) plane. Complicated geometries exhibiting much complex structure may best be handled using a composite grid.

The boundary-fitted system approach also offers a substantial advantage over the usual treatment of curved boundaries with finite differences. For example, a rectangular mesh may have been overlayed on any general shape and either a "saw tooth" approximation or interpolation could be employed for the curved regions. It is reasonable to expect that the boundary-fitted coordinate system would provide a better boundary representation in such areas. Other advantages of the boundary-fitted approach include:

(1) Simplification of boundary condition specification;

(2) accurate implementation of boundary conditions;

(3) efficient use of grid points for the calculation area; and

(4) flexibility in the type of curved-wall problems that can be addressed.

Once the physical domain is specified, a suitable boundary-fitted computing mesh can be generated. Unfortunately, this is seldom an easy task. The reason for this arises partly from the difficulty in providing a sufficiently good initial estimate for the grid and in the choice of an appropriate numerical procedure. Thompson and Warsi [67] provide a comprehensive review of numerical methods for generating curvilinear coordinate systems having coordinate lines coincident with boundary segments.

8.2 Mathematical Modelling

Modelling pulsatile flow through the circulatory system in precise mathematical terms remains an insurmountable problem. The system is affected by many physical and chemical factors. In the case of prosthetic heart valves, the situation is further complicated by the presence of an unnatural device in the flow path. In spite of the difficulty created by the multiplicity of interrelated elements, extensive insight into fluid flows can still be obtained by applying numerical solution techniques to simpler physical systems.

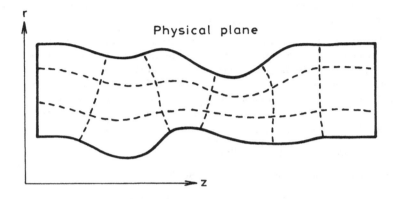

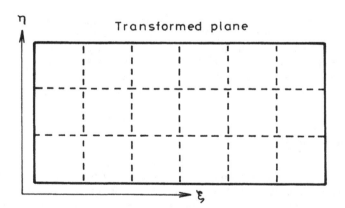

Fig. 8.1: Example of a non-rectangular geometry in the physical plane with schematic grid lines and its corresponding representation in the transformed plane.

The literature contains many examples of mathematical models for a variety of systems (physical, biological, chemical, etc ...) with analytic or numerical methods of solution available. Consequently, this chapter is not concerned in detailing the many complex equations which make up fluid mechanics but instead restricts the discussion to a relatively qualitative level. This chapter describes some basic fluid laws relating to flow through blood vessels and prosthetic heart valves. Mathematical models of turbulence are discussed. Computational methods are covered for solving nonlinear mathematical flow equations and generating boundary-fitted coordinate systems.

8.3 Laminar Versus Turbulent Flow Models

As we have seen earlier, there are two basic modes of viscous fluid motion — laminar and turbulent flow. If we consider axisymmetric flows in a pipe, then in laminar flow particles follow straight lines parallel to the pipe axis, but in turbulent flow the particles follow a random path through the pipe, and only the average motion is along the pipe axis. Laminar and turbulent flows are essentially states of stability and instability, respectively, and as such, no complete definition of turbulence seems possible.

As explained earlier, experimental observations indicate that above a critical Reynolds number, transition will occur from laminar to turbulent flow. Thus, laminar flow can become unstable. Below a certain Reynolds number, small disturbances introduced to the flow are always damped out and the flow remains stable. The random fluctuating motion characteristic of turbulent flow arises from small disturbances which are not damped out, but instead increase with time and eventually interact with each other in a chaotic manner. As a result, differences between laminar and turbulent flow become evident. Firstly, the velocity profiles appear very different. The fully developed laminar profile is parabolic and the turbulent profile is much flatter near the pipe centre with larger gradients near the vessel walls. Secondly, the "roughness" of turbulent flow acts to increase the effective viscosity of the fluid. In the laminar case, the viscosity is entirely due to molecular effects. However, for turbulent flow the apparent viscosity comes from two sources:

(i) molecular viscosity; and

(ii) convection of turbulent "eddies" which usually form the larger part of the effective viscosity. This additional term is often referred to as the *eddy viscosity*.

In the circulatory system, the flow of blood is normally laminar. However, abrupt flow variations due to irregularities and stenoses, as are often encountered with prosthetic heart valves, will produce turbulence. Experimental research indicates that turbulent flow can occur during the period of ejection in the ascending aorta of subjects with elevated cardiac output, and it occurs consistently in the ascending aorta of subjects with abnormal aortic valves or with prosthetic aortic valves (Stein and Sabbah, [61]). It is apparent from the discussion above that turbulence will cause a substantial increase in the level of shear stress since both the effective viscosity and the shear strain rate will become larger, particularly near a vessel wall. Given valve geometry, fluid properties and details of the inflow conditions one would prefer to evaluate in totality the velocity and shear stress fields for the entire flow domain.

The connection between the observed complexity of fluid flow and the fundamental physical laws governing such behaviour is usually obscure. Indeed, with what we know, both the basic laws and experimental observation are required to make the connection. Unfortunately, a mathematical description of fluid flow based on physical laws has not yet developed sufficiently so as to cater for every possible flow situation. Even in cases where the mathematics is known, the formulation is often posed in highly nonlinear terms, thus excluding analytic solution. Similarly, the implementation of experimental measurements is time-consuming and expensive, and the accuracy of observations, as limited as they may be, is often questioned.

Nevertheless, the need to address real-life problems is pressing, and mathematical models of fluid flow, both laminar and turbulent, have been developed and applied successfully to many problems. These models have governing partial differential equations which are based on the well known conservation principles of mass, momentum and energy. The mathematical formulation is usually nonlinear and analytic application to realistic problems is often very difficult or impossible unless simplifications are made. By exploiting the capabilities of modern high-speed computers and numerical techniques, solutions to otherwise unsolvable problems are being found.

The following two sections provide a discussion of turbulence models and the numerical methods that have been applied to fluid flow problems.

8.4 Turbulence Models

Turbulence modelling of complex engineering and physical flows has been a topic of active research for many years. Several mathematical models have been developed, and these are usually classified as either Reynolds stress models or eddy viscosity models. Reynolds stress models are based on the evolutionary equations for the turbulent stresses. They were developed to overcome deficiencies in eddy viscosity models. Although successful in computing free shear flows, they have encountered difficulties in treating wall-bounded separated flows. In addition, their complex structure results in increased computer time and decreased stability. It is questionable whether the benefits outweigh the costs. As a result, Reynolds stress models have been applied to relatively few turbulent flow problems.

Eddy viscosity models, however, are much more popular. These models are based on the *Kolmogorov-Prandtl* proposal that the eddy viscosity is proportional to the product of a turbulent velocity scale and a turbulent length scale. That is

$$\nu_t \propto qL \tag{8.1}$$

where ν_t is the eddy viscosity, q and L are velocity and length scales of the turbulence, respectively. Depending upon the manner in which the eddy viscosity is specified, these models can be further grouped into either algebraic or differential models. With algebraic models ν_t is determined by replacing q and L with simple algebraic expressions involving the mean velocity field. Mixing-length models, as they are also known, are excellent for investigating single-zone flows but become deficient when switching between one type of zone to another within a single flow. This is particularly noticeable when investigating separated flows.

Differential models specify the eddy viscosity in the solution of one or two differential equations. One-equation models prescribe the turbulent length scale but use a partial differential equation to determine the turbulent kinetic energy. The kinetic energy serves to characterise the velocity scale. This approach results in only a marginal improvement over the mixing-length models.

Two-equation models employ the Boussinesq eddy-viscosity concept

$$\overline{u_i'u_j'} = -\nu_t \left(\frac{\partial U_i}{\partial x_j} + \frac{\partial U_j}{\partial x_i}\right) + \frac{2}{3}k\delta_{ij} \tag{8.2}$$

where δ_{ij} is the Kronecker delta, and U_i and U_j are the mean components of velocity in the i and j directions, respectively. Differential equations are used to involve both the velocity and length scales. One of the most studied two-equation turbulence models is the $k-\epsilon$ *model*. This model remains the most widely used approach for the solution of practical turbulent flow problems. In this model, the eddy viscosity is obtained by solving partial differential transport equations for the turbulent kinetic energy k and the energy dissipation rate ϵ. These transport equations are relatively easy to incorporate into existing Navier-Stokes computer codes. The complete equations are presented in the subsequent sections.

The $k-\epsilon$ model has been successfully applied to a great number of shear flow problems. However, for more complicated flows the model tends to lose its accuracy. The standard $k-\epsilon$ model is based on the assumption that the eddy viscosity is identical for all Reynolds stresses. In two-dimensional flows without swirl this condition has proved adequate for the calculation of thin shear layers. However, in certain three-dimensional flow situations (eg., square duct flows) and in planar separated flows the isotropic eddy viscosity assumption proves insufficient by underpredicting velocity fluctuations, and thus giving rise to inaccurate predictions for the Reynolds stresses. The normal Reynolds stresses play an important role in the calculation of the mean velocity field in separated turbulent flows. For example, it is well known that the $k - \epsilon$ model underpredicts the recirculation length downstream of a backward-facing step. Another consequence is the inability to predict secondary flows in non-circular ducts. A refinement which uses an algebraic stress model can be introduced to replace both the Kolmogorov-Prandtl expression (8.1) and the Boussinesq eddy-viscosity relation (8.2) above (Rodi [55]).

A limitation on the $k - \epsilon$ turbulence model is the restriction to high Reynolds number flows (ie., $R_e > 20000$). Consequently, it is not directly applicable to the viscous sublayer near boundary walls and to low Reynolds number flows. However, the common approach of using wall functions allows efficient treatment of the near wall regions. In the case of low Reynolds number flows the standard $k - \epsilon$ model and wall-function method are insufficient and alternative approaches have been developed. The most popular of these is the low-Reynolds number turbulence model of Jones and Launder [32]. This model seeks to provide turbulence equations which are valid throughout the laminar, semi-laminar and fully turbulent regions. The costs associated with this improvement are the requirement for many additional computations in the near wall region, increased equation complexity and related numerical instability.

Nevertheless, the $k - \epsilon$ turbulence model and wall-function method is the simplest approach being widely applied to practical turbulent flow problems. Once implemented, it is the stepping stone by which the model refinements mentioned above may eventually be introduced.

8.5 Computational Methods for the Study of Flow Through Prosthetic Heart Valves

As explained in Chapter 7, the *in vitro* measurement of flow through heart valve prostheses produces useful but limited information, and in some instances, such as the potential for valve thrombosis, very little can be deduced from a finite number of measuring points. As a result, extensive clinical trials are required to establish whether prosthetic valves have acceptably low levels of blood trauma and thrombosis. Computational fluid dynamic techniques, however, provide a powerful tool for determining valve haemodynamic performance as they can yield detailed quantitative data for velocity and stress distributions in the flow field. Thus, conclusions regarding the potential for a valve to induce blood trauma or give rise to thrombotic deposition can be drawn via computational methods. Additional benefits include reliability, efficiency, immediacy and ready adaption to various geometric and flow configurations. Nevertheless, computational techniques and experimental measurements will continue to complement each other as they currently do in many aspects of modelling in other physical sciences.

Several authors have employed numerical methods to determine blood flow patterns in the vicinity of prosthetic heart valves. Au and Greenfield [1] applied numerical techniques and principles of fluid dynamics to study stress distribution in blood caused by the axial motion of an occluder in a caged-disk prosthesis. They assumed planar geometry, laminar Newtonian flow and a comparatively low Reynolds number of 50 (based upon distance between the occluder and cage strut, and the occluder velocity). Time-dependent finite difference equations were formulated and an interactive graphics environment provided continuous monitoring of numerical procedure for stability and convergence rates. Changes in maximum stress values as the occluder moved from the fully closed to an almost fully open position were reported. However, the emphasis was on the application of interactive computer graphics to study this and other medical phenomena.

Au and Greenfield [2] again utilised finite difference methods to examine steady laminar flow through a planar model of the Björk-Shiley tilting-disk prosthesis. The accent was again on the totality of the graphics displays which included contour maps and surface plots of computer-formed streamline, velocity, stress and pressure fields. Potential sites for haemodynamically induced thrombogenesis were observed. Importantly, the difficulty associated with computing on irregular boundary segments was noted and it was suggested that the finite element approach could overcome this problem.

Underwood and Mueller [68] obtained numerical solutions for steady axisymmetric laminar flow through a disk-type valve in a constant diameter chamber for Reynolds numbers ranging from 20 to 1300. The finite difference method was used to compute flow variables and the solution procedure includes calculation of a suitable time-step at each interval. Experimentation with both first-order and second-order accurate upwind differences for the advection terms revealed variation in the numerical results — the second method provides better solutions when compared with physical measurements. Regions of separated flow attached to the disk occluder, downstream of the sewing ring and along the aortic wall were identified. Zones of high fluid stress were also noted. The possibility of thrombosis and haemolysis was discussed.

In an extension of the above study, Underwood and Mueller [69] repeated the calculations, but instead of assuming a constant diameter chamber a more realistic aortic-shaped vessel was considered. A non-uniform grid system was used for both the axial and radial directions enabling close approximation of the aortic wall shape. Differences in solutions between the two chambers were reported. Reynolds numbers examined were between 50 and 600. Regions of separated flow were again present, particularly a recirculating zone on the chamber wall downstream of the sinus of Valsalva. This region was initially observed at a Reynolds number of about 200 and grew with increasing flow rates.

Engelman *et. al.* [14] performed comparative studies on the steady laminar flow characteristics of Starr-Edwards caged-ball and Björk-Shiley tilting-disk valves in the mitral and aortic positions. The finite element method was used to model planar non-Newtonian fluid flow through the prostheses. The location of major eddies were identified in velocity vector plots for both valve types. Simulations using a Newtonian model were also carried out with markedly different results to the non-Newtonian model being reported for each configuration case.

Peskin [49] developed a computational method whereby the mathematical problem is formulated in such a way that the heart muscle and valves appear as special regions of the flow where extra forces apply. These forces give the valves and tissue walls their material and physiological properties. He assumed laminar Newtonian time-dependent flow. The flow equations were coupled to a nonlinear system of equations describing boundary motion and forces. Unfortunately, the method was restricted to very low Reynolds numbers ($\approx 1/25$ of the appropriate value) due to impractical computational demands. The study illustrated the use of computational methods to optimise prosthetic heart valve design and encouraged the use of the method to examine valve performance when subjected to changes in the physiological state of the heart (eg., exercise).

Yang and Wang [71,72] applied the $k - \epsilon$ turbulence model to investigate the problem of steady turbulent flow through a disk-type prosthetic heart valve in a constant diameter chamber. They predicted streamline, velocity, stress and pressure fields by solving a set of simultaneous algebraic finite difference equations using a line-iteration technique. Unlike laminar flow studies which can have up to four separated flow regions in a constant diameter chamber and three separated regions in an aortic-shaped chamber, the turbulent study uncovered only two such regions. Specifically, attached to the downstream side of the sewing ring and the disk occluder — there was no recirculation zone present along the chamber wall. High levels of fluid stress and wall shear stress were also evident.

Idelsohn *et. al.* [30] carried out a comparative study on steady laminar flow through three common prosthetic heart valves. Using the finite element method, velocity, stress and pressure fields were calculated for caged-ball, caged-disk and tilting-disk prostheses in an aortic-shaped chamber. Regions of accelerated and stagnant flow were presented with the aid of computer graphics. Also, the location and magnitude of maximum shear stresses were found. Although these positions were generally in good agreement with experimental measurements, the levels of stress were much lower than those observed in physical simulators. The discrepancy between the numerical and experimental values indicated the important contribution of turbulent shear stresses.

Stevenson and Yoganathan [62,63] conducted a numerical investigation for steady turbulent flow through axisymmetric models of trileaflet tissue aortic heart valves. Body conforming grid systems were generated and the turbulent flow equations were solved using finite differences and successive over-relaxation. They computed the flow patterns for five valve geometries ranging from very stenotic to a natural heart valve model. Good agreement was reported between experimental measurements and simulation data for velocity, shear stress and pressure drops.

Their numerical model provided much useful information and may become a valuable tool for the development of future tissue prostheses.

8.6 Laminar Flow Model Through a Prosthesis

Let us consider a model for steady laminar flow through the idealized planar geometry of a disk-type valve. Although very simple, this model clearly demonstrates the potential effectiveness and usefulness of more sophisticated numerical and mathematical models. The method of finite differences is used to solve the model. Numerical solutions are provided for several Reynolds numbers.

8.6.1 *Problem Formulation*

The mathematical modelling of blood flow through the complex geometry of a prosthetic heart valve is a difficult task. In such a problem the complex geometries of the valve must be simplified so that they can be studied numerically. The simplifications must still contain the essential features of the valve. An approximate solution to the realistic problem may then be deduced from the study performed on the simplified model with the necessary assumptions taken into account. Thus, an insight into the real-life problem can be obtained.

The present analysis is performed on a disk-type prosthetic heart valve which can be used in either the mitral or aortic position, see Mazumdar and Thalassoudis [41]. In this study the valve is assumed to be in the aortic position. The sinuses of Valsalva and the structure of the valve cage will undoubtedly influence the flow field near an aortic valve. However, the occluder and the sewing ring will remain the overwhelming factors influencing the flow characteristics of the valve. Thus, it is decided not to include the sinuses and the valve cage in the present simple model. For the purpose of mathematical modelling, the laminar incompressible two-dimensional steady flow of a homogeneous Newtonian fluid with constant viscosity is assumed. The flow is considered during the greater part of systole when the valve is fully open. This is the worst case in terms of the extent of flow separation and the magnitude of fluid stresses.

A simple geometry is employed for which experimental data is available. The essential features present in this model are the sewing ring, the disk occluder, the aortic wall and a Poiseuille inflow velocity profile. The flow is assumed to be symmetric about the centreline, and hence only one half of the actual geometry is used for the analysis. A definition sketch for the disk valve model is given in Fig. 8.2. The dimensions of the valve are given in Fig. 8.3.

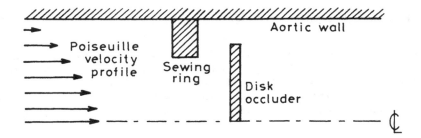

Fig. 8.2: Geometry and inflow velocity profile for the disk valve.

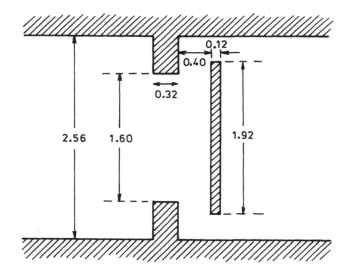

Fig. 8.3: Dimensions (cm) of the fully open disk valve.

Based upon the foregoing assumptions, the governing equations for two-dimensional steady incompressible flow of a Newtonian fluid with no body forces are the continuity equation and the Navier-Stokes equations (ie., Eqs. 1.52-1.53)

$$\frac{\partial u}{\partial x} + \frac{\partial v}{\partial y} = 0, \tag{8.3}$$

$$u\frac{\partial u}{\partial x} + v\frac{\partial u}{\partial y} + \frac{\partial p}{\partial x} = \frac{1}{R_e}\left(\frac{\partial^2 u}{\partial x^2} + \frac{\partial^2 u}{\partial y^2}\right), \tag{8.4}$$

$$u\frac{\partial v}{\partial x} + v\frac{\partial v}{\partial y} + \frac{\partial p}{\partial y} = \frac{1}{R_e}\left(\frac{\partial^2 v}{\partial x^2} + \frac{\partial^2 v}{\partial y^2}\right) \tag{8.5}$$

where u and v are the horizontal and vertical velocity components, respectively, and p is the pressure. The Reynolds number R_e is defined as

$$R_e = \frac{D\overline{U}}{\nu} \tag{8.6}$$

where $\overline{U}$ is the average inflow velocity, D is the channel width and ν is the kinematic viscosity of the fluid.

By introducing the stream function $\psi(x,y)$ and the vorticity function $\omega(x,y)$ as defined in Chapter 1

$$u = \frac{\partial \psi}{\partial y} \quad \text{and} \quad v = -\frac{\partial \psi}{\partial x} \tag{8.7}$$

and

$$\omega = \frac{1}{2}\left(\frac{\partial v}{\partial x} - \frac{\partial u}{\partial y}\right) \tag{8.8}$$

we finally obtain the vorticity-transport equation and Poisson's equation,

$$\frac{\partial \psi}{\partial y}\frac{\partial \omega}{\partial x} - \frac{\partial \psi}{\partial x}\frac{\partial \omega}{\partial y} = \frac{1}{R_e}\left(\frac{\partial^2 \omega}{\partial x^2} + \frac{\partial^2 \omega}{\partial y^2}\right), \tag{8.9}$$

$$\frac{\partial^2 \psi}{\partial x^2} + \frac{\partial^2 \psi}{\partial y^2} = -2\omega. \tag{8.10}$$

The nonlinearity of the coupled partial differential equations and the disk valve geometry make it impossible to obtain a solution through an analytic approach. Instead, the use of finite difference techniques provides a way by which an approximate numerical solution for the problem may be obtained. Once a numerical solution becomes available the velocity components can be calculated using a discretisation of Eq. 8.7. The shear stress defined by

$$\sigma_{xy} = -\frac{1}{R_e}\left(\frac{\partial u}{\partial y} + \frac{\partial v}{\partial x}\right) \tag{8.11}$$

can then be obtained through a further discretisation, as can the horizontal normal stress

$$\sigma_{xx} = -\frac{2}{R_e}\left(\frac{\partial u}{\partial x}\right). \qquad (8.12)$$

The boundary conditions imposed are:

(1) *Upstream boundary*

A fully developed Poiseuille velocity profile for channel flow is assumed at the inflow boundary. Flow in large arteries is found to assume such profiles. To ensure a horizontal inflow, the vertical velocity component v is set to zero. The inflow stream function and vorticity are subsequently computed from Eqs. 8.7 and 8.8.

(2) *Centreline*

The stream function is given an arbitrary value of zero along the centreline. The values of the vorticity and the vertical velocity component are also equated to zero along this axis of symmetry.

(3) *Solid walls*

The stream function is given a constant (non-zero) value along the entire solid boundary which includes the sewing ring and the aortic wall. Along the occluder boundary the stream function is equated to zero. The no-slip boundary conditions ensure that both velocity components are zero in these locations. The vorticity on solid walls is calculated from Eqs. 8.7 and 8.8 by taking a Taylor series expansion of the stream function out from the wall.

(4) *Corner points*

The value of the stream function at corner points deserves no special treatment and is taken as a constant value. The vorticity at a convex corner point is calculated as an average of two values obtained by considering the x and y directions separately. The vorticity in concave corner points is equated to zero.

(5) *Downstream boundary*

The outflow boundary conditions employed by Friedman [17] for the stream function and vorticity are also used here. These conditions are

$$\frac{\partial \psi}{\partial x} = 0 \quad \text{and} \quad \frac{\partial \omega}{\partial y} - R_e \frac{\partial \psi}{\partial y}\left(\omega + \frac{1}{2}\frac{\partial^2 \psi}{\partial y^2}\right) = 0 \qquad (8.13)$$

which satisfy the requirements for horizontal outflow and constant pressure far downstream. The distance of the outflow boundary conditions from the disk valve is considered to be sufficiently far so as not to severely influence the flow upstream.

8.6.2 *Finite Difference Formulation*

Before the mathematical model can be formulated into a set of finite difference equations a computational mesh must be constructed. For simplicity a regularly spaced mesh is chosen and boundary locations are made to coincide with the nearest grid point. The dimensions of the grid spacing in x and y are denoted by Δx and Δy, respectively, and are both equal to 0.04 cm. The computational mesh is a 164×33 grid (5412 grid points). During the initial code development and testing several other grid densities were investigated. Comparison between the numerical solutions indicated some ambiguous results occurring at the lower grid densities. These grids were discarded. A section of the final grid system is shown in Fig. 8.4.

The governing Eqs. 8.9 and 8.10 are now written in finite difference form. There are numerous ways in which this discretisation can be achieved. For this simple model the method employed by Friedman [17] is used as the basis for the finite difference formulation. Defining H and K for convenience as

$$H = \psi_{i+1,j} - \psi_{i-1,j}, \qquad K = \psi_{i,j+1} - \psi_{i,j-1} \tag{8.14}$$

the advection terms in the vorticity transport equation (8.9) are approximated using first order accurate upwind differences if $H \geq 0$ and $K \geq 0$. That is,

$$\left.\frac{\partial \omega}{\partial x}\right|_{i,j} \approx \frac{\omega_{i,j} - \omega_{i-1,j}}{\Delta x} \tag{8.15}$$

and

$$\left.\frac{\partial \omega}{\partial y}\right|_{i,j} \approx \frac{\omega_{i,j+1} - \omega_{i,j}}{\Delta y} \tag{8.16}$$

if $H \geq 0$ and $K \geq 0$.

The remaining first derivatives of ψ and second derivatives representing the diffusion terms are approximated by the well known central space approximations. Substitution into Eq. 8.9 and rearranging yields the following finite difference equation

$$\omega_{i,j} = \frac{\omega_{i+1,j} + \omega_{i-1,j} + \omega_{i,j+1} + \omega_{i,j-1} + \frac{R_e}{2}(H\omega_{i,j+1} + K\omega_{i-1,j})}{4 + \frac{R_e}{2}(H + K)} \tag{8.17}$$

if $H \geq 0$ and $K \geq 0$. Similar finite difference equations are formulated for the other three possibilities, namely

$$\omega_{i,j} = \frac{\omega_{i+1,j} + \omega_{i-1,j} + \omega_{i,j+1} + \omega_{i,j-1} + \frac{R_e}{2}(H\omega_{i,j+1} - K\omega_{i+1,j})}{4 + \frac{R_e}{2}(H + K)} \tag{8.18}$$

if $H \geq 0$ and $K < 0$,

$$\omega_{i,j} = \frac{\omega_{i+1,j} + \omega_{i-1,j} + \omega_{i,j+1} + \omega_{i,j-1} + \frac{R_e}{2}(K\omega_{i-1,j} - H\omega_{i,j-1})}{4 + \frac{R_e}{2}(K - H)} \tag{8.19}$$

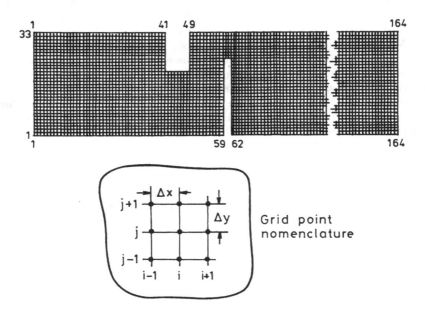

Fig. 8.4: Finite difference grid (164 × 33) and grid nomenclature for the disk valve.

if $H < 0$ and $K \geq 0$, and

$$\omega_{i,j} = \frac{\omega_{i+1,j} + \omega_{i-1,j} + \omega_{i,j+1} + \omega_{i,j-1} - \frac{R_e}{2}(K\omega_{i+1,j} + H\omega_{i,j-1})}{4 - \frac{R_e}{2}(K + H)} \tag{8.20}$$

if $H < 0$ and $K < 0$. The presence of R_e in both the numerator and denominator in the equations above overcomes the cell Reynolds number restriction often associated with solving such fluid flow problems.

Boundary conditions are also discretised. A difficulty of any computational viscous fluid flow problem which employs the stream function-vorticity method is that the values of vorticity on the no-slip solid boundaries are not known in advance but are required to solve the discretised equations. To overcome this difficulty Eq. 8.8

is used to provide an initial estimate for the wall vorticity once the initial values for the stream function are specified. The wall vorticity at the grid point (i, j), denoted by $\omega_{i,j}$, is given by the second order accurate expression

$$\omega_{i,j} \approx \frac{3(\psi_{i,j} - \psi_{i+1,j})}{(\Delta x)^2} - \frac{\omega_{i+1,j}}{2} \tag{8.21}$$

when the solid wall is facing in the positive x-direction. Similar expressions are used for the other possible configurations.

8.6.3 *Numerical Solution*

The numerical procedure employed by Friedman provides a way by which a numerical solution to a steady laminar flow at high Reynolds numbers can be obtained using a relatively small amount of computer time. The procedure is used in a step by step manner to calculate both the stream function and the vorticity as follows:

(i) Set initial values for $\psi^{(1)}$ and $\omega^{(1)}$.

(ii) Use Successive Over Relaxation (SOR) with a relaxation factor of 1.8 only once to calculate the stream function at all the grid points immediately interior to the solid boundaries using the new vorticity values.

(iii) Calculate the stream function at all remaining interior grid points using the new values determined in Step 2 and the most recent vorticity values.

(iv) Return to Step 2 and repeat the process until a convergent solution is obtained. Denote the convergent result by $\overline{\psi}^{(k+1)}$.

(v) Set
$$\psi^{(k+1)} = 0.96\,\overline{\psi}^{(k+1)} + 0.04\,\psi^{(k)}. \tag{8.22}$$

(vi) Use the Gauss-Seidel iteration method only once to calculate the boundary vorticity values using the new stream function values.

(vii) Calculate the vorticity at all interior grid points using the new values determined in Step 6 and the latest stream function values.

(viii) Return to Step 6 and repeat the process until a convergent solution is obtained. Denote the convergent result by $\overline{\omega}^{(k+1)}$.

(ix) Set
$$\omega^{(k+1)} = 0.3\,\overline{\omega}^{(k+1)} + 0.7\,\omega^{(k)}. \tag{8.23}$$

(x)   Return to Step 2 and repeat the process until both

$$\left| \psi^{(k+1)} - \psi^{(k)} \right| < \epsilon_1 \qquad (8.24)$$

and

$$\left| \omega^{(k+1)} - \omega^{(k)} \right| < \epsilon_2 \qquad (8.25)$$

are satisfied.

The stream function was iterated using the SOR method until the difference between its value at consecutive iterate levels was less than 0.0005. The vorticity function was iterated with the Gauss-Seidel method until the difference between its value at consecutive iterate levels was less than 0.2% of the maximum vorticity value. A requirement was that the convergence criteria be satisfied at every grid point. The global convergence conditions were set by assigning values to ϵ_1 and ϵ_2 of 0.0005 and 0.2% of the maximum vorticity value, respectively. Final global convergence was declared when Eqs. 8.24 and 8.25 were both satisfied 30 times.

Convergent numerical solutions are obtained for Reynolds numbers $R_e = 50$, 500, 1000 and 6000. Stream function, vorticity, horizontal velocity, vertical velocity and shear stress solutions are computed at every grid point. The numerical solutions presented are of course subject to artificial viscosity errors introduced by the numerical procedure. This should be kept in mind when comparisons are made with experimental studies, particularly in regions of high velocity diagonal to grid lines.

Stream function contour plots are presented in Fig. 8.5 for $R_e = 50$ to 6000. The extent of recirculating regions downstream of the disk occluder as well as upstream and downstream of the sewing ring are clearly seen. The location of the separation point on the occluder remained at the downstream corner for all Reynolds numbers tested. The large recirculating zone behind the occluder contained horizontal reverse flow velocities up to 1.57 times ($i = 80, j = 1$) the average inflow velocity for $R_e = 6000$. Interesting features in Fig. 8.5 are the relative dimensions of the closed streamlines in the separated flow region immediately downstream of the occluder. As the Reynolds number increases from 500 to 6000, the streamlines, $\psi = -0.05, -0.10$ and -0.15 become markedly smaller, until the one at the innermost ($\psi = -0.05$) eventually vanishes for $R_e > 1000$. Physically, the flow velocity inside the separated zone decreases relative to the surrounding flow as the Reynolds number increases.

A recirculating region of large proportions also appeared along the aortic wall downstream of the occluder (Fig. 8.5). Although it was not present at $R_e = 50$, this region initially appeared at about $R_e = 150$ and grew rapidly with higher Reynolds numbers. The separation point for this region was observed to move upstream as the Reynolds number increased. The largest value of wall shear stress occurring within this zone was -4.1 Nm^{-2} ($i = 80, j = 33$) for $R_e = 6000$.

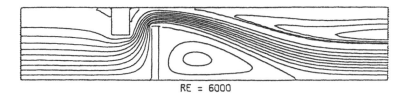

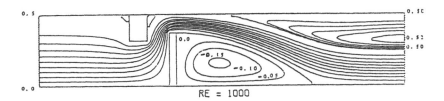

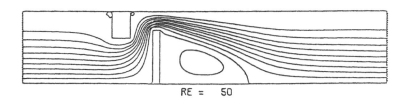

Fig. 8.5: Computed streamlines for R_e=50, 1000 and 6000.

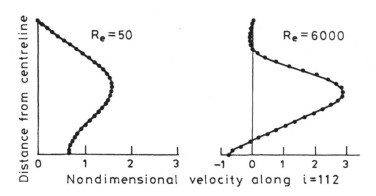

(a) $i = 112$

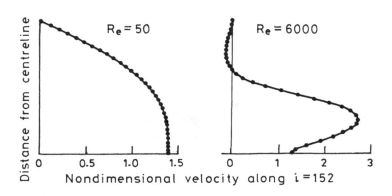

(b) $i = 152$

Fig. 8.6: Computed horizontal velocity profiles for $R_e = 50$ and $R_e = 6000$ along various grid lines.

In Fig. 8.6 plots of the computed horizontal velocity profiles at various cross sections of the flow field are shown. The changing structure of the horizontal velocity profiles at different locations downstream of the valve can be seen as the Reynolds number increases from 50 to 6000. They show the presence of the slow moving reversed flow region lying along the aortic wall. The potential for thrombus formation in this stagnated flow area is clearly enhanced, particularly if the endothelium has been damaged. It is interesting to note that whereas the $R_e = 50$ horizontal velocity profile in Fig. 8.6b almost re-estalished the parabolic profile it had at inflow, the same cannot be said for the $R_e = 6000$ case. The great flow disturbance that occurs at this high Reynolds number persists far downstream.

The distribution of wall shear stress along the aortic wall is given in Fig. 8.7 for $R_e = 50$ and 6000. Large variations in the wall shear stress occur as a result of lateral flow (induced by the valve geometry) impinging upon the vessel boundary. The "sharp" corners of the sewing ring structure also influence the wall shear stress levels. Points of flow separation are indicated in Fig. 8.7 by zero shear stress. The maximum levels of wall shear stress computed upstream of the sewing ring remained below 0.7 Nm^{-2} for $R_e = 6000$. The location of peak wall shear stress remained fixed ($i = 61, j = 33$) for all Reynolds numbers. The shear stress at this point was 61.8 Nm^{-2} for $R_e = 6000$. This is sufficient to damage endothelial cells, adhered red blood cells and platelets. However, the point of peak wall stress coincides with the location of the sinuses of Valsalva which are not considered in this simple model. In reality, the presence of these hollowed-out areas in the aortic root region would reduce the amount of stress, possibly averting extensive damage. The levels of wall shear stress diminished rapidly downstream from this point for all Reynolds numbers. The level of shear stress computed along the sewing ring reached a maximum value of 44.4 Nm^{-2} ($i = 47, j = 21$) for $R_e = 6000$. The maximum level of shear stress in the entire flow domain occurred just below the upstream corner of the occluder reaching 82.8 Nm^{-2} ($i = 59, j = 23$) for $R_e = 6000$.

The laminar shear stresses computed in this model appear likely to cause damage to adhered red blood cells, endothelial cells and platelets in several regions of the flow domain. Damage to red blood cells in flow is unlikely to occur based on the present results. However, when flow becomes turbulent, the turbulent stresses which are always much larger than viscous stresses could haemolyse blood cells in the vicinity of solid surfaces.

The potential for thrombus formation exists near the regions of separated flow observed at all flow rates. Thrombosis may occur in the recirculating region immediately behind the occluder, attaching itself to the restrictive cage, possibly inhibiting the mechanical performance of the disk valve. Thromboplastin may also be released from blood cells in the separated region lying alongside the aortic wall to enhance clotting. These conditions are clearly undesirable.

Although these results are obtained quite easily from a simple laminar model, they do provide a good illustration of the detailed descriptions of shear stress and

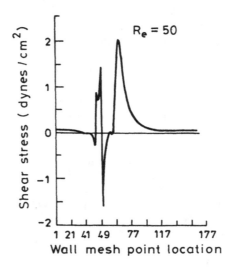

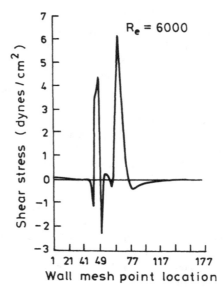

Fig. 8.7: Computed wall shear stress distributions (dynes cm^{-2}) for $R_e = 50$ and $R_e = 6000$.

velocity that are possible in the near vicinity of prosthetic heart valves. Both the flow patterns and the stress distributions indicate areas of concern and potential danger. However, it should be emphasised that turbulence will occur at the higher Reynolds numbers. The subsequent models incorporate the effects of turbulence.

8.7 Turbulent Flow Model Through a Ball Prosthesis

8.7.1 *Introduction*

As an extension of the previous model, let us consider another model which examines steady turbulent axisymmetric flow through a Starr-Edwards caged-ball valve in an aortic-shaped chamber. A boundary-fitted coordinate system is generated for the flow domain in the vicinity of the valve. The coordinate lines follow the left ventricular wall, the sewing ring, and idealized sinus, the aortic wall and the occluder shape. The standard $k - \epsilon$ turbulence model is formulated using a stream function-vorticity approach. The governing partial differential equations are recast into their curvilinear equivalents before being discretised into finite difference equations which are then solved iteratively (Thalassoudis *et. al.* [66]). The main aims of this turbulent model are to obtain quantitative estimates for velocity, wall shear stress and Reynolds stress distributions in the flow field surrounding the prosthetic valve.

8.7.2 *Model Formulation*

The geometry representing a fully open Starr-Edwards 1260-11A caged-ball prosthetic heart valve in an aortic-shaped chamber is shown in Fig. 8.8. The geometry includes a left ventricular chamber, a sewing ring structure, and idealized axisymmetric sinus, the aortic wall and a ball occluder. The dimensions of the valve are given in Fig. 8.9. For comparison purposes the dimensions are very close to those employed by Figliola and Mueller [15] in their experimental measurements. Assumptions made are that blood behaves like a Newtonian fluid in large arteries, the cage containing the occluder has a negligible effect upon the flow, the domain boundaries are smooth and rigid and the flow is axisymmetric. The fluid is given a density of 1.06 gm cm^{-3} and a molecular, or laminar, viscosity of 3.5×10^{-3} N sec m^{-2} closely matching that of human blood.

The analysis of turbulent flow is more difficult than the case of the laminar flow model discussed before, and it has become customary to abandon the study of the detailed flow structure and treat the flow as consisting of mean and fluctuating components. Using this approach the governing mass and momentum equations are decomposed through application of time-averaging. This procedure results in additional terms $\rho\overline{u_i'u_j'}$ often referred to as turbulent, or Reynolds stresses where ρ is the fluid density, u_i' and u_j' are the fluctuating components of velocity in the i and j directions, respectively. Closure of the model equations is achieved through the Boussinesq eddy-viscosity concept which assumes that the turbulent stresses are directly proportional to the mean velocity gradients. In general form this is expressed by

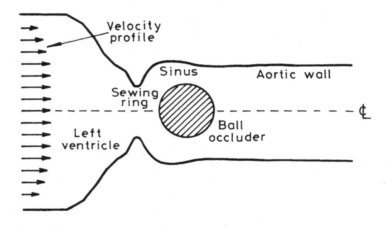

Fig. 8.8: Geometry and inflow velocity profile for the ball valve model.

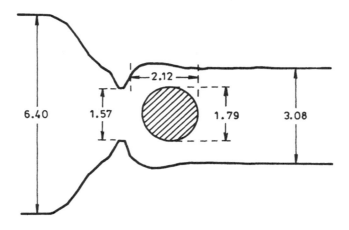

Fig. 8.9: Dimensions (cm) of the fully open ball valve.

$$\overline{u_i'u_j'} \;=\; -\nu_t \left(\frac{\partial U_i}{\partial x_j} + \frac{\partial U_j}{\partial x_i} \right) + \frac{2}{3} k \delta_{ij} \tag{8.26}$$

where δ_{ij} is the Kronecker delta, U_i and U_j are the mean components of velocity in the i and j directions, respectively, and ν_t is the turbulent kinematic viscosity defined by

$$\nu_t \;=\; C_\mu \frac{k^2}{\epsilon} \tag{8.27}$$

where k is the turbulent kinematic energy representing the intensity of the turbulent fluctuations, ϵ is the dissipation rate of turbulent kinetic energy representing the length scale of the energy containing eddies and C_μ is an empirical constant.

The resulting partial differential equations together with the additional terms required for closure are referred to as the $k - \epsilon$ turbulence model. For this model the governing equations are written in a stream function-vorticity ($\psi - \omega$) formulation. Thus, we have the following equations (see Yang and Wang [72])

$$\omega \;=\; -\frac{1}{r}\left(\psi_{zz} + \psi_{rr} - \frac{1}{r}\psi_r \right), \tag{8.28}$$

$$\frac{1}{r}\left(\psi_r\omega_z - \psi_z\omega_r + \frac{\omega}{r}\psi_z\right) = (\nu_{eff}\omega)_{zz} + (\nu_{eff}\omega)_{rr} + \frac{1}{r}(\nu_{eff}\omega)_r - \frac{1}{r^2}\nu_{eff}\omega \quad (8.29)$$

where the effective kinematic viscosity ν_{eff} is defined as

$$\nu_{eff} = \nu + \nu_t \quad (8.30)$$

where ν is the molecular kinematic viscosity. The additional equations for the turbulent kinetic energy and its dissipation rate for a steady axisymmetric flow necessary to complete the model are

$$\frac{1}{r}(\psi_r k_z - \psi_z k_r) = \left[\left(\frac{\nu_{eff}}{\sigma_k}k_z\right)_z + \frac{1}{r}\left(r\frac{\nu_{eff}}{\sigma_k}k_r\right)_r\right] + G - C_D\epsilon, \quad (8.31)$$

$$\frac{1}{r}(\psi_r\epsilon_z - \psi_z\epsilon_r) = \left[\left(\frac{\nu_{eff}}{\sigma_\epsilon}\epsilon_z\right)_z + \frac{1}{r}\left(r\frac{\nu_{eff}}{\sigma_\epsilon}\epsilon_r\right)_r\right] + \frac{\epsilon}{k}(C_1 G - C_2\epsilon). \quad (8.32)$$

The generation of turbulent kinetic energy G from the mean flow is given by

$$G = \nu_t\left\{2\left[U_z^2 + V_r^2 + \left(\frac{V}{r}\right)^2\right] + (U_r + V_z)^2\right\} \quad (8.33)$$

where U and V, the mean axial and radial velocities, respectively, are defined by

$$U = \frac{1}{r}\psi_r \quad \text{and} \quad V = -\frac{1}{r}\psi_z. \quad (8.34)$$

$C_1, C_2, C_D, C_\mu, \sigma_k$ and σ_ϵ are empirical constants at high Reynolds numbers (Launder and Spalding [37]) whose values are given by

$$C_1 = 1.44, \quad C_2 = 1.92, \quad C_D = 1.0, \quad C_\mu = 0.09, \quad \sigma_k = 1.0, \quad \sigma_\epsilon = 1.3.$$

The governing equations are subject to several boundary conditions. At inflow we assume a fully developed turbulent velocity profile for pipe flow given by (Schlichting [57])

$$U = U_0\left(1 - \frac{r}{R}\right)^{\frac{1}{7}}, \quad V = 0 \quad (8.35)$$

where U_0 is the mean axial centreline velocity at inflow, and R the inflow chamber radius. Analytic expressions for the inflow stream function and vorticity can be obtained from Eq. 8.35.

The turbulent kinetic energy and its dissipation rate at inflow are approximated by

$$k = 0.03\,U^2, \quad (8.36)$$

$$\epsilon = \frac{C_D k^{\frac{3}{2}}}{\lambda} \qquad (8.37)$$

where λ is the macro eddy length scale. The numerical solutions proved insensitive to differing inflow conditions for ψ, ω, k and ϵ with the large blocking effect of the valve dominating the flow conditions in the vicinity of the prosthesis.

At outflow the boundary condition $\Phi_z = 0$ ($\Phi = \psi, \omega, k, \epsilon$) is imposed, and along the centreline the conditions are $\Phi_r = 0$ ($\Phi = k, \epsilon$), $\psi = 0$ and $\omega = 0$. The position of the outflow boundary at a distance of 10 R past the ball occluder ensured the outflow conditions had little effect upon the flow upstream. Along solid boundaries the usual no-slip conditions are applied (ie., $U = 0$ and $V = 0$) and the values of k and ϵ are also set to zero. The usual 'law of the wall' conditions are specified on the first grid point away from the wall. That is

$$\frac{U_p}{U^*} = \frac{1}{\kappa} \ln\left(Ey^+\right), \qquad (8.38)$$

$$y^+ = \frac{yU^*}{\nu}, \qquad (8.39)$$

$$k = \frac{(U^*)^2}{\sqrt{C_\mu}}, \qquad (8.40)$$

$$\epsilon = \frac{(U^*)^3}{\kappa y}, \qquad (8.41)$$

$$\omega = \frac{U^*}{\kappa y} \qquad (8.42)$$

where U_p is the velocity parallel to the wall, U^* is the friction velocity, κ is von Karman's constant ($\kappa = 0.4187$), E is van Driest's constant ($E = 9.793$), y^+ is the local Reynolds number and y is the grid point distance from the wall.

The velocity parallel to the wall at the first grid point away from the wall is computed along with the grid point distance from the boundary. An initial estimate for the friction velocity is made by assuming $U^* \approx |U_p|$. Based upon this estimate, y^+ is then evaluated. If $y^+ \leq 0.15$, the values of k and ϵ are equated to zero and the value of ω calculated from Eq. 8.29 is used. This approach is effective in controlling the stability of the numerical scheme during the early iterations, especially in the vicinity of separation and reattachment points where y^+ would occasionally assume very low values. For the vast majority of grid points adjacent to the boundary U^* is determined through a simple iterative process. If it is found that $y^+ > 11.63$, ie., turbulence effects dominate, then k, ϵ and ω are computed from the equations above. Otherwise, if $y^+ \leq 11.63$ the flow is assumed to be purely laminar and the friction velocity is given by $U^* = y^+$ before k, ϵ and ω are computed. The sign of the vorticity computed above is assigned depending upon the sign of the corresponding U_p.

All the governing partial differential equations are recast into curvilinear coordinates by substituting derivatives in the (z, r) physical plane by their equivalent expressions in the (ξ, η) transformed plane.

Equations 8.28 and 8.29 are transformed into (ξ, η) coordinates to obtain

$$\omega = -\frac{1}{r}\left[\frac{1}{J^2}(\alpha\psi_{\xi\xi} - 2\beta\psi_{\xi\eta} + \gamma\psi_{\eta\eta} + \sigma\psi_\eta + \tau\psi_\xi) - \frac{1}{Jr}(z_\xi\psi_\eta - z_\eta\psi_\xi)\right] \tag{8.43}$$

and

$$\frac{1}{Jr}\left[\psi_\eta\omega_\xi - \psi_\xi\omega_\eta + \frac{\omega}{r}(r_\eta\psi_\xi - r_\xi\psi_\eta)\right]$$

$$= \frac{1}{J^2}\{\alpha(\nu_{eff}\omega)_{\xi\xi} - 2\beta(\nu_{eff}\omega)_{\xi\eta}$$

$$+ \gamma(\nu_{eff}\omega)_{\eta\eta} + \sigma(\nu_{eff}\omega)_\eta + \tau(\nu_{eff}\omega)_\xi\}$$

$$+ \frac{1}{Jr}\{z_\xi(\nu_{eff}\omega)_\eta - z_\eta(\nu_{eff}\omega)_\xi\} - \frac{\nu_{eff}\omega}{r^2} \tag{8.44}$$

where

$$J = z_\xi r_\eta - z_\eta r_\xi, \quad \alpha = z_\eta^2 + r_\eta^2, \quad \beta = z_\xi z_\eta + r_\xi r_\eta, \quad \gamma = z_\xi^2 + r_\xi^2, \tag{8.45}$$

$$\sigma = \frac{r_\xi(\alpha z_{\xi\xi} - 2\beta z_{\xi\eta} + \gamma z_{\eta\eta}) - z_\xi(\alpha r_{\xi\xi} - 2\beta r_{\xi\eta} + \gamma r_{\eta\eta})}{J} \tag{8.46}$$

and

$$\tau = \frac{z_\eta(\alpha r_{\xi\xi} - 2\beta r_{\xi\eta} + \gamma r_{\eta\eta}) - r_\eta(\alpha z_{\xi\xi} - 2\beta z_{\xi\eta} + \gamma z_{\eta\eta})}{J}. \tag{8.47}$$

For an orthogonal coordinate system $\beta = 0$.

We further have,

$$\phi_z \equiv \left(\frac{\partial\phi}{\partial z}\right)_{r,t} = (r_\eta\phi_\xi - r_\xi\phi_\eta)/J, \tag{8.48}$$

$$\phi_r \equiv \left(\frac{\partial\phi}{\partial r}\right)_{z,t} = (z_\xi\phi_\eta - z_\eta\phi_\xi)/J, \tag{8.49}$$

$$\phi_{zz} \equiv \left(\frac{\partial^2\phi}{\partial z^2}\right)_{r,t} = (r_\eta^2\phi_{\xi\xi} - 2r_\xi r_\eta\phi_{\xi\eta} + r_\xi^2\phi_{\eta\eta})/J^2$$

$$+ \left[(r_\eta^2 r_{\xi\xi} - 2r_\xi r_\eta r_{\xi\eta} + r_\xi^2 r_{\eta\eta})(z_\eta\phi_\xi - z_\xi\phi_\eta)\right.$$

$$\left. + (r_\eta^2 z_{\xi\xi} - 2r_\xi r_\eta z_{\xi\eta} + r_\xi^2 z_{\eta\eta})(r_\xi\phi_\eta - r_\eta\phi_\xi)\right]/J^3, \tag{8.50}$$

$$\phi_{rr} \equiv \left(\frac{\partial^2 \phi}{\partial r^2}\right)_{z,t} = (z_\eta^2 \phi_{\xi\xi} - 2z_\xi z_\eta \phi_{\xi\eta} + z_\xi^2 \phi_{\eta\eta})/J^2$$

$$+ \left[(z_\eta^2 r_{\xi\xi} - 2z_\xi z_\eta r_{\xi\eta} + z_\xi^2 r_{\eta\eta})(z_\eta \phi_\xi - z_\xi \phi_\eta) \right.$$

$$\left. + (z_\eta^2 z_{\xi\xi} - 2z_\xi z_\eta z_{\xi\eta} + z_\xi^2 z_{\eta\eta})(r_\xi \phi_\eta - r_\eta \phi_\xi)\right]/J^3,$$

$$(8.51)$$

$$\phi_{zr} = \left[(z_\xi r_\eta + z_\eta r_\xi)\phi_{\xi\eta} - z_\xi r_\xi \phi_{\eta\eta} - z_\eta r_\eta \phi_{\xi\xi}\right]/J^2$$

$$+ \left[z_\eta r_\eta z_{\xi\xi} - (z_\xi r_\eta + z_\eta r_\xi)z_{\xi\eta} + z_\xi r_\xi z_{\eta\eta}\right](r_\eta \phi_\xi - r_\xi \phi_\eta)/J^3$$

$$+ \left[z_\eta r_\eta r_{\xi\xi} - (z_\xi r_\eta + z_\eta r_\xi)r_{\xi\eta} + z_\xi r_\xi r_{\eta\eta}\right](z_\xi \phi_\eta - z_\eta \phi_\xi)/J^3$$

$$(8.52)$$

and

$$\nabla^2 \phi = (\alpha \phi_{\xi\xi} - 2\beta \phi_{\xi\eta} + \gamma \phi_{\eta\eta} + \sigma \phi_\eta + \tau \phi_\xi)/J^2, \qquad (8.53)$$

$$\vec{\nabla}\phi = [(r_\eta \phi_\xi - r_\xi \phi_\eta)\vec{i} + (z_\xi \phi_\eta - z_\eta \phi_\xi)\vec{j}]/J, \qquad (8.54)$$

$$\frac{1}{Jr}(\psi_\eta k_\xi - \psi_\xi k_\eta) = \left(\frac{\nu_{eff}}{\sigma_k}\right)\left[\frac{1}{J^2}(\alpha k_{\xi\xi} - 2\beta k_{\xi\eta} + \gamma k_{\eta\eta}\right.$$

$$\left. + \sigma k_\eta + \tau k_\xi) + \frac{1}{Jr}(z_\xi k_\eta - z_\eta k_\xi)\right]$$

$$+ \frac{1}{J^2}(r_\eta k_\xi - r_\xi k_\eta)\left\{r_\eta \left(\frac{\nu_{eff}}{\sigma_k}\right)_\xi - r_\xi \left(\frac{\nu_{eff}}{\sigma_k}\right)_\eta\right\}$$

$$+ \frac{1}{J^2}(z_\xi k_\eta - z_\eta k_\xi)\left\{z_\xi \left(\frac{\nu_{eff}}{\sigma_k}\right)_\eta - z_\eta \left(\frac{\nu_{eff}}{\sigma_k}\right)_\xi\right\}$$

$$+ G - C_D \epsilon \qquad (8.55)$$

and

$$\frac{1}{Jr}(\psi_\eta \epsilon_\xi - \psi_\xi \epsilon_\eta) = \left(\frac{\nu_{eff}}{\sigma_\epsilon}\right)\left[\frac{1}{J^2}(\alpha \epsilon_{\xi\xi} - 2\beta \epsilon_{\xi\eta} + \gamma \epsilon_{\eta\eta}\right.$$

$$\left. + \sigma \epsilon_\eta + \tau \epsilon_\xi) + \frac{1}{Jr}(z_\xi \epsilon_\eta - z_\eta \epsilon_\xi)\right]$$

$$+ \frac{1}{J^2}(r_\eta \epsilon_\xi - r_\xi \epsilon_\eta)\left\{r_\eta \left(\frac{\nu_{eff}}{\sigma_\epsilon}\right)_\xi - r_\xi \left(\frac{\nu_{eff}}{\sigma_\epsilon}\right)_\eta\right\}$$

$$+ \frac{1}{J^2}(z_\xi \epsilon_\eta - z_\eta \epsilon_\xi) \left\{ z_\xi \left(\frac{\nu_{eff}}{\sigma_\epsilon} \right)_\eta - z_\eta \left(\frac{\nu_{eff}}{\sigma_\epsilon} \right)_\xi \right\}$$

$$+ \frac{\epsilon}{k}(C_1 G - C_2 \epsilon). \tag{8.56}$$

The generation of turbulent kinetic energy G term given by Eq. 8.33 is also transformed to (ξ, η) coordinates.

Finally the Reynolds number for this model is defined by

$$R_e = \frac{DU_0}{\nu} \tag{8.57}$$

where D is the inflow chamber diameter and U_0 is the mean axial centreline velocity at inflow.

8.7.3 *Coordinate System Generation*

In the previous section, the geometric configuration of the caged-disk model is simple and the boundary walls are coincident with the Cartesian coordinate system. In this section, details are presented on the approach used to generate an orthogonal boundary-fitted coordinate system for a fully open Starr-Edwards caged-ball prosthesis.

The method requires that the shape of physical domain be known in advance and not be determined as part of the solution. Initially, the upper surface of the boundary shape shown in Fig. 8.8 was constructed from a combination of trigonometric functions for the left ventricular chamber, sinus of Valsalva and aortic wall, and cubic splines for the sewing ring. A hemisphere was used for the ball occluder (as only half the valve geometry is necessary due to the symmetry assumption) but this was altered slightly by "smoothing" the fore and aft sections of the hemisphere to reduce singularity effects which became evident during the computation process. The boundary functions for the upper part of the physical domain are:

(1) Inflow chamber, $0.0 \leq z < 4.0$:

$$r(z) = 3.2. \tag{8.58}$$

(2) Left ventricular section, $4.0 \leq z < 5.8$:

$$r(z) = 0.95 \cos \left[\frac{\pi}{2}(z - 4) \right] + 2.25. \tag{8.59}$$

(3) Sewing ring, $5.8 \leq z < 6.7$:

$$r_i(z) = \sum_{j=1}^{4} c_{ij}(z - z_i)^{j-1}, \quad z_i \leq z < z_{i+1}, \quad i = 1, \ldots, 7 \tag{8.60}$$

where the values of z_i and coefficients c_{ij} are given in Table 8.1.

(4) Sinus of Valsalva, $6.7 \leq z < 10.0$:

$$r(z) = 0.03 \, (10 - z)^{1.5} \sin \left[\frac{3\pi}{4}(z - 6) - \frac{\pi}{2} \right] + 1.54. \tag{8.61}$$

(5) Aortic wall, $10.0 \leq z \leq 25.0$:

$$r(z) = 1.54. \tag{8.62}$$

Table 8.1: Cubic spline coefficients for the sewing ring structure.

i	z_i	c_{i1}	c_{i2}	c_{i3}	c_{i4}
1	5.8000	1.345	−0.41715	0.00000	−43.28503
2	5.9000	1.260	−1.71570	−12.98551	61.42514
3	6.0000	1.020	−2.47005	5.44203	5.41959
4	6.0800	0.860	−1.49527	6.74274	0.39943
5	6.2225	0.785	0.45075	6.91349	0.89021
6	6.3600	0.980	2.40245	7.28071	−42.32405
7	6.4800	1.300	2.32142	−7.95595	12.05447
8	6.7000	1.554	0.57111	0.00000	0.22000

The approach adopted for generating a boundary-fitted grid is that described by Mobley and Stewart [45]. The goal is to evaluate grid point values in the physical plane (z, r) subject to the requirement that the grid lines in the transformed plane (ξ, η) form a conjugate harmonic pair, thereby causing the $\xi - \eta$ contours to be orthogonal. That is, they must satisfy the Cauchy-Riemann equations

$$\xi_z = \eta_r, \tag{8.63}$$

$$\xi_r = -\eta_z. \tag{8.64}$$

In order to control the separation of grid lines in the physical plane, ξ and η are defined as monotonically increasing functions of two variables χ and ζ, respectively. That is,

$$\xi = f(\chi), \quad \eta = g(\zeta). \tag{8.65}$$

Laplace-type equations are formulated for each of the spatial coordinates in the (z, r) physical plane. In general form, the Laplace-type equations are

$$\mathcal{L}(z) = 0, \tag{8.66a}$$

$$\mathcal{L}(r) = 0 \tag{8.66b}$$

where the operator $\mathcal{L}$ is defined as

$$\mathcal{L} = \frac{\partial^2}{\partial \chi^2} + \delta^2 \frac{\partial^2}{\partial \zeta^2} - F \frac{\partial}{\partial \chi} - \delta^2 G \frac{\partial}{\partial \zeta} \tag{8.67}$$

and

$$\delta(\chi, \zeta) \equiv \frac{f'(\chi)}{g'(\zeta)}, \quad F(\chi) \equiv \frac{f'(\chi)}{f''(\chi)}, \quad G(\zeta) \equiv \frac{g'(\zeta)}{g''(\zeta)} \tag{8.68}$$

where $f' \equiv df/d\chi$, $f'' \equiv d^2 f/d\chi^2$, etc., and $f' \neq 0$, $g' \neq 0$.

To complete orthogonality the solutions for both z and r are coupled at the domain boundaries by application of the Cauchy-Riemann conditions. A weak constraint is applied so that the final distribution of boundary grid points is not predetermined but is computed through the numerical mapping procedure. For example, in order to solve Eq. 8.66a, z must also satisfy the Neumann condition

$$z_\zeta = -\frac{g'}{f'} r_\chi \tag{8.69}$$

along the top and bottom boundaries, whilst Dirichlet conditions are specified on the other domain boundaries. Similar conditions are applied to the domain boundaries when solving Eq. 8.66b, except that Dirichlet conditions are applied to the top boundary using Eqs. 8.58-8.62, to the bottom boundary using the patched hemisphere equation, and a Neumann condition on the remaining walls.

A finite difference method employing central space derivative approximations was developed for solving Eqs. 8.66a and 8.66b but the resultant grid was noticeably non-orthogonal, particularly near boundary segments with high surface gradients. Conventional error analysis revealed that the coefficients of higher ordered terms in the vicinity of the steep boundary segments were assuming extremely large values. This problem was partially resolved by switching to either backward or forward differencing (depending upon the orientation of the boundary segment) when computing in these regions.

A further improvement was obtained by implementing a simple procedure described by Davies [13] where orthogonal trajectories are used for the orthogonalisation of a discrete non-orthogonal coordinate system. An iterative scheme is used for the numerical solution of the grid equations. An adequate initial guess is obtained by simple linear interpolation of the boundary values in the radial direction, and equidistant grid lines in the axial direction (see Thalassoudis, Mazumdar *et. al.* [43]). The "overlapping" of grid lines predominant during early iterations stems largely from the rather poor initial guess and the accompanying complexity of the geometric configuration. Several other initial estimates for z and r involving compression and interpolation of grid lines were attempted, but with little success. After

300 iterations the (ξ, η) coordinate lines were no longer crossing. Convergence, albeit non-orthogonal, was declared after 500 iterations. These values of z and r were then employed as initial values for orthogonalisation using the method of Davies [13]. Excess grid lines in the radial direction were removed and some additional lines were eventually inserted in the axial direction. The final computational grid (120×16) is shown in Fig. 8.10.

It should be remembered that the production of orthogonal meshes is by no means an easy task as it depends heavily upon domain geometry. From a practical point of view, it would be more efficient to produce a general coordinate system.

8.7.4 *Finite Difference Formulation*

The transformed governing equations are discretised in the (ξ, η) coordinate system. Central differencing is applied to all derivative terms in Eq. 8.43 and rearranged to yield

$$
\psi_{i,j} = \left\{ 2 \left(\frac{\alpha}{\Delta\xi^2} + \frac{\gamma}{\Delta\eta^2} \right) \right\}^{-1} \times \left[\alpha \frac{(\psi_{i+1,j} + \psi_{i-1,j})}{(\Delta\xi)^2} \right.
$$

$$
- 2\beta \frac{(\psi_{i+1,j+1} - \psi_{i+1,j-1} + \psi_{i-1,j-1} - \psi_{i-1,j+1})}{4(\Delta\xi)(\Delta\eta)}
$$

$$
+ \gamma \frac{(\psi_{i,j+1} + \psi_{i,j-1})}{(\Delta\eta)^2} + \sigma \frac{(\psi_{i,j+1} - \psi_{i,j-1})}{2(\Delta\eta)}
$$

$$
+ \tau \frac{(\psi_{i+1,j} - \psi_{i-1,j})}{2(\Delta\xi)} - \frac{J}{r} \left(\frac{(z_{i+1,j} - z_{i-1,j})}{2(\Delta\xi)} \frac{(\psi_{i,j+1} - \psi_{i,j-1})}{2(\Delta\eta)} \right.
$$

$$
\left. \left. - \frac{(z_{i,j+1} - z_{i,j-1})}{2(\Delta\eta)} \frac{(\psi_{i+1,j} - \psi_{i-1,j})}{2(\Delta\xi)} \right) + J^2 \omega_{i,j} r \right] \tag{8.70}
$$

where the indices (i, j) refer to a computational grid point and $(\Delta\xi)$, $(\Delta\eta)$ are the grid spacings in the (ξ, η) system.

Equation 8.44 is discretised in a similar manner, except for terms such as ω_ξ and ω_η which are treated using upwind differencing for stability purposes. That is, if u and v are the components of velocity in the ξ and η directions, respectively, defined by

$$
u = \frac{z_\xi U + r_\xi V}{\sqrt{\gamma}}, \qquad v = -\frac{z_\eta U + r_\eta V}{\sqrt{\alpha}} \tag{8.71}
$$

then

$$
\omega_\xi \approx \begin{cases} \dfrac{\omega_{i,j} - \omega_{i-1,j}}{(\Delta\xi)}, & \text{if } u \geq 0 \\[2mm] \dfrac{\omega_{i+1,j} - \omega_{i,j}}{(\Delta\xi)}, & \text{if } u < 0 \end{cases} \tag{8.72}
$$

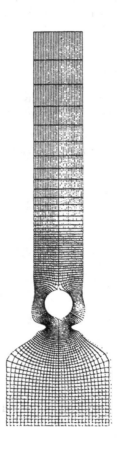

Fig. 8.10: The final computational grid (120 × 16).

and

$$\omega_\eta \approx \begin{cases} \dfrac{\omega_{i,j} - \omega_{i,j-1}}{(\Delta\eta)}, & \text{if } v \geq 0 \\ \dfrac{\omega_{i,j+1} - \omega_{i,j}}{(\Delta\eta)}, & \text{if } v < 0. \end{cases} \qquad (8.73)$$

The one-sided approximations for the vorticity together with the central space approximations for the remaining derivatives are substituted into Eq. 8.44 and re-arranged to yield a finite difference equation for $\omega_{i,j}$. Upwind differencing is applied in a similar manner to obtain the k and ϵ finite difference forms from Eqs. 8.55 and 8.56, respectively, but are not reproduced here for brevity.

8.7.5 *Numerical Solution*

The finite difference equations are solved iteratively with relaxation parameters to control stability and convergence. Alternating the direction of the iterative sweeps also enhances stability. The inflow boundary conditions for the ψ, k and ϵ are extended throughout the flow domain as initial conditions. Vorticity values along the wall are at first approximated from the initial stream function values with vorticity values being set to zero elsewhere. Once a numerical solution is obtained it is used as the initial condition for a higher Reynolds number flow.

The iterative scheme employed is summarized in steps as follows:

(i)   Specify initial values for $\psi^{(1)}, \omega^{(1)}, k^{(1)}$ and $\epsilon^{(1)}$.

(ii) Update the stream function at grid points immediately away from solid boundaries.

(iii) Iterate the stream function at interior grid points with a relaxation factor of 1.75 until convergence. Denote the result $\overline{\psi}^{(n+1)}$.

(iv) Set
$$\psi^{(n+1)} = 0.96\,\overline{\psi}^{(n+1)} + 0.4\,\psi^{(n)}. \qquad (8.74)$$

(v)   Evaluate U, V and G.

(vi) Update the boundary vorticity values.

(vii) Iterate the vorticity at interior grid points with a relaxation factor of 0.9 until convergence. Denote the result $\overline{\omega}^{(n+1)}$.

(viii) Set
$$\omega^{(n+1)} = 0.3\,\overline{\omega}^{(n+1)} + 0.7\,\omega^{(n)}. \qquad (8.75)$$

(ix) Evaluate ν_{eff}, U, V and G.

(x)   Apply 'law of the wall' to grid points immediately away from solid boundaries.

(xi)   Evaluate the turbulent kinetic energy at interior points with a relaxation parameter of 0.15. Denote the result $\overline{k}^{(n+1)}$.

(xii)   Set

$$k^{(n+1)} = 0.3\,\overline{k}^{(n+1)} + 0.7\,k^{(n)}. \qquad (8.76)$$

(xiii) Apply 'law of the wall'.

(xiv) Evaluate the turbulent kinetic energy dissipation rate at interior points with a relaxation parameter of 0.15. Denote the result $\overline{\epsilon}^{(n+1)}$.

(xv)   Set

$$\epsilon^{(n+1)} = 0.3\,\overline{\epsilon}^{(n+1)} + 0.7\,\epsilon^{(n)}. \qquad (8.77)$$

(xvi) Test for global convergence and return to Step (ii) if necessary.

Global convergence is declared when all dependent variables change by less than 1% of their maximum value between iteration steps for at least 50 times.

The accuracy of the model numerical solutions was verified by repeating the calculations for $R_e = 1000$, 2000 and 4000 on a 120×31 coordinate system. This was necessary to ensure that a sufficiently dense grid had been chosen for the computations.

Converged numerical solutions for steady turbulent flow through a fully open Starr-Edwards 1260-11A prosthetic heart valve are obtained for Reynolds numbers of 1000 and 6000 corresponding to steady flow rates of 8.133 and 48.796 litres min^{-1}, respectively.

Stream function contours are presented in Fig. 8.11. The solutions show two regions of separated flow. A large separation zone is attached to the downstream side of the ball occluder and a smaller annular zone is attached to the sewing ring and inside the sinus. No separated regions are predicted upstream of the sewing ring nor along the aortic wall. The experimental studies conducted in [15] on flow through a Starr-Edwards 1260-11A caged-ball valve in an aortic-shaped chamber contained separated flow regions attached to the sewing ring (upstream and downstream), the aortic side of the cage struts and distal to the ball occluder. The separation region immediately upstream of the sewing ring was only described qualitatively and appeared to be very small.

The computed mean axial velocity profiles are given in Fig. 8.12. At the $z = 30$ mm location, the computed velocity profiles appear to be partly influenced by the narrowing chamber dimensions further upstream. They no longer conform to the fully developed turbulent velocity profiles specified at inflow.

Reverse flow (-38 cm sec^{-1}) present in the sinus separation zone can be clearly seen in the $z = 70$ mm velocity profile. The maximum axial velocity (187 cm sec^{-1}) occurs very close to the upstream surface of the ball occluder. This profile is very similar in shape to the one experimentally measured at an identical location.

Computed wall shear stress profiles are shown in Fig. 8.13 for $R_e = 4000$. Upstream of the sewing ring the wall shear stress remains below 3 Nm^{-2}. The shear stress increases rapidly as the flow encounters the primary valve orifice reaching a maximum of 138.3 Nm^{-2} at the sewing ring tip. The wall shear decreases rapidly downstream of the sewing ring but elevated levels are again observed along the sinus boundary. Along this boundary the wall shear stress attains a negative peak of -8.1 Nm^{-2} which is a direct consequence of the velocities in the sinus separation zone. Further downstream, a broad region of elevated wall stress is seen along the aortic boundary corresponding to where the flow impinges upon the wall. The peak level of computed aortic wall shear stress is 30.7 Nm^{-2}. The maximum computed stress along the occluder wall was -63.2 Nm^{-2} for $R_e = 4000$. The level of stress rises and falls rapidly on either side of this peak.

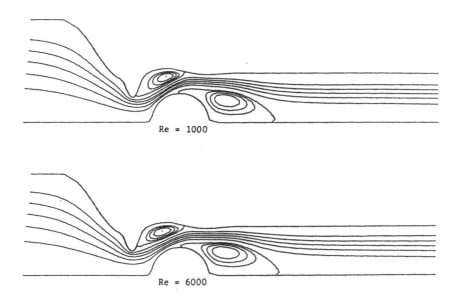

Re = 1000

Re = 6000

Fig. 8.11: Computed mean streamlines for $R_e = 1000$ and 6000.

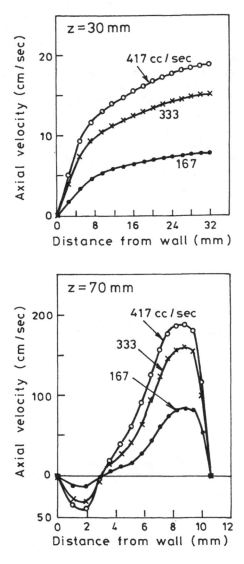

Fig. 8.12: Computed mean axial velocity profiles at 30 mm and 70 mm from the inflow boundary for a Starr-Edwards ball valve.

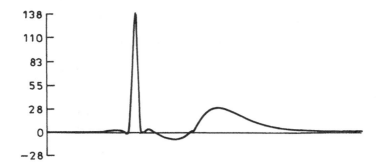

(a) Computed aortic wall shear stress (Nm^{-2});

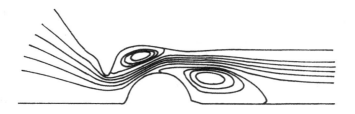

(b) corresponding streamlines; and

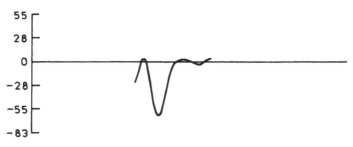

(c) occluder wall shear stress (Nm^{-2}) for $R_e = 4000$.

Fig. 8.13.

8.7.6 *Conclusion*

The numerical models discussed in this section using computational fluid mechanics provide detailed quantitative descriptions for the steady flow characteristics of caged-disk and caged-ball prosthetic heart valves. The first model, is very simple. It considers laminar flow through a caged-disk valve at several Reynolds numbers. The results indicate the potential effectiveness of such computational studies, but the model is limited in application because of the laminar flow assumption. The second model considers turbulent flow through a caged-ball prosthesis in an aortic-shaped chamber. A boundary-fitted coordinate system is generated and the governing flow equations are conveniently solved on a rectangular transformation plane. The behaviour of the model is generally consistent with experimental observations. Regions of slow moving flow, high fluid stress and elevated wall shear are identified.

This study presents solutions to a numerical model for steady turbulent flow through a Starr-Edwards caged-ball prosthesis in an aortic-shaped chamber. This valve has an estimated 122,000 worldwide implants, thus underlining the practical importance of this research.

Using computational fluid mechanics techniques to solve complex mathematical models has provided valuable and detailed quantitative data for velocities and stress fields in the vicinity of prosthetic heart valves — a totality currently unattainable through physical simulators.

Ghista [22] pointed out the "unlimited potential" of computer design and graphics for the biomechanical design of prostheses and surgical procedures. Accordingly, the long term application of this form of research is seen both as an effective design tool and to assist cardiac surgeons in the selection of a suitable valve implant for individual patients. The application of mathematical modelling and computational fluid mechanics for the evaluation of the flow characteristics of prosthetic heart valves has and will continue to provide much valuable information.

BIBLIOGRAPHY

The list of references included here is meant to be a helpful guide through the literature rather than a complete catalogue. My sincere apologies to the many authors whose work is not in the list. The references are listed alphabetically according to the first author. There are several related references, which are ordered according to date of publication.

1. A.D. Au and H.S. Greenfield, *Comput. Biol. Med.* **4** (1975) 279.

2. A.D. Au and H.S. Greenfield, *Comput. Biomed. Res.* **10** (1977) 165.

3. C.P. Bailey, *Dis. Chest* **15** (1949) 377.

4. D.W. Baker and R.E. Daigle, in *Cardiovascular Flow Dynamics and Measurements*, ed. N.H.C. Hwang and N.A. Normann (University Park Press, Baltimore, 1977).

5. B.J. Bellhouse and L. Talbot, *J. Fluid Mech.* **35** (1969) 721.

6. D.H. Bergel, *Cardiovascular Fluid Dynamics, Vol. 2* (Academic Press, London and New York, 1972) p. 158.

7. V.O. Björk, *Scand. J. Thorac. Cardiovasc. Surg.* **3** (1969) 1.

8. P.L. Blackshear, in *Chemistry of Biosurfaces, Vol. 2*, ed. M.L. Hair (Marcel Dekker, New York, 1972) p. 523.

9. J.J. Blum, *Am. J. Physiol.* **198** (1960) 991.

10. E. Buckingham, *Phys. Rev.* **4** (1914) 345.

11. M. Casson, in *Rheology of Dispersive Systems*, ed. C.C. Mills (Pergamon, Oxford, 1959) p. 84.

12. S.E. Charm and G.S. Kurland, *Nature* **206** (1965) 617.

13. C.W. Davies, *J. Comput. Phys.* **39** (1981) 164.

14. M.S. Engelman, S.E. Moskowitz and J.B. Borman, *J. Thorac. Cardiovasc. Surg.* **79** (1980) 402.

15. R.S. Figliola and T.J. Muller, *A.S.M.E. J. Biomech. Eng.* **99** (1977) 173.

16. J.H. Forrester and D.F. Young, *J. Biomech.* **3** (1970) 297.

17. M. Friedman, *J. Eng. Math.* **6** (1972) 285.

18. D.L. Fry, *Circ. Res.* **22** (1968) 165.

19. Y.C. Fung, *Biomechanics: Mechanical Properties of Living Tissues* (Springer-Verlag, New York, 1981).

20. Y.C. Fung, *Biodynamics Circulation* (Springer-Verlag, New York, 1984).

21. Y.C. Fung, *Biomechanics: Motion, Flow, Stress and Growth* (Springer-Verlag, New York, 1990).

22. D.N. Ghista, *Proc. 2nd Int. Conf. Mech. Med. Biol.* (Osaka, Japan, 5-7 June, 1980).

23. G. Hagen, *Poggendorffs Ann.* **46** (1839) 423.

24. D.E. Harken, H.S. Soroff, W.J. Taylor, A.A. Lefemine, S.K. Gupta and S. Lunzer, *J. Thorac. Cardiovasc. Surg.* **40** (1960) 744.

25. J.D. Hellums and C.H. Brown, in *Cardiovascular Fluid Dynamics*, ed. N.H.C. Hwang and N.A. Normann (University Park Press, Baltimore, 1977) p. 799.

26. W.H. Herkes and J.R. Lloyd, *A.S.M.E. J. Biomech. Eng.* **103** (1981) 267.

27. C.A. Hufnagel, W.P. Harvey, P.J. Rabil and T.F. McDermott, *Surgery* **35** (1954) 673.

28. C.A. Hufnagel and P.W. Conrad, *Surgery* **57** (1965) 205.

29. T.C. Hung, R.M. Hochmuch, J.H. Joist and S.P. Sutera, *Trans. Am. Soc. Artif. Intern. Organs* **22** (1976) 285.

30. S.R. Idelsohn, L.E. Costa and R. Ponso, *J. Biomech.* **18** (1985) 97.

31. G.B. Jeffery, *Proc. Roy. Soc. A.* **102** (1922) 161.

32. W.P. Jones and B.E. Launder, *Int. J. Heat Mass Transfer* **16** (1973) 119.

33. J.N. Kapur, *Mathematical Models in Biology and Medicine* (Affiliated East-West Press PVT. Ltd., India, 1985).

34. R.L. Kaster and C.W. Lillehei, *Dig. 7th Int. Conf. Med. Biol. Eng.* (Stockholm, 1967) p. 387.

35. E.B. Kay, A. Suzuki, M. Demaney and H.A. Zimmerman, *Amer. J. Cardiol.* **18** (1966) 504.

36. A. Krogh, *J. Phys.* **52** (1919) 457.

37. B.E. Launder and D.B. Spalding, *Comput. Methods Appl. Mech. Eng.* **3** (1974) 269.

38. L.B. Leverett, J.D. Hellums, C.P. Alfrey and E.C. Lynch, *Biophys. J.* **12** (1972) 257.

39. J. Lighthill, *Physiological Fluid Mechanics* (Springer-Verlag, New York, 1971).

40. J. Lighthill, *Mathematical Biofluid Dynamics* (SIAM, Philadelphia, U.S.A., 1975).

41. J. Mazumdar and K. Thalassoudis, *Med. Biol. Eng. Comput.* **21** (1983) 400.

42. J. Mazumdar, *An Introduction to Mathematical Physiology and Biology* (Cambridge University Press, Cambridge, 1989).

43. J. Mazumdar, K.C. Ang and L.L. Soh, *J. Australas. Phys. Eng. Sci. Med.* **14** (1991) 65.

44. E.W. Merrill, A.M. Benis, E.R. Gilliland, T.K. Sherwood and E.W. Salzman, *J. Appl. Physiol.* **20** (1965) 954.

45. C.D. Mobley and R.J. Stewart, *J. Comput. Phys.* **34** (1980) 124.

46. B.E. Morgan and D.F. Young, *Bull. Math. Biol.* **36** (1974) 39.

47. C.G. Nevaril, J.D. Hellums, C.P. Alfrey Jr. and E.C. Lynch, *AIChE Journal* **15** (1969) 707.

48. B.J. Noye, *Proc. Int. Conf. Numer. Simul. Fluid Dyn. Syst.* (Monash University, Melbourne, Australia, 1976) p. 1.

49. C.S. Peskin, *J. Comput. Phys.* **25** (1977) 220.

50. J. Poiseuille, *Compt. Rend.* **11** (1840) 961, 1041.

51. J. Poiseuille, *Compt. Rend.* **12** (1841) 112.

52. H. Reul, M. Giersiepen and E. Knott, in *Heart Valve Engineering* (MEP, London, U.K., 1986).

53. M.R. Roach and A.C. Burton, *Can. J. Biochem. Physiol.* **35** (1957) 681.

54. P.J. Roache, *Computational Fluid Dynamics* (Hermosa, Albuquerque, New Mexico, 1976).

55. W. Rodi, *Turbulence Models and their Application to Hydraulics — A state of the art review* (International Association for Hydraulic Research, Delft, 1980).

56. S.I. Rubinow and J.B. Keller, *J. Theor. Biol.* **35** (1972) 299.

57. H. Schlichting, *Boundary Layer Theory*, (6th ed.) (McGraw-Hill, New York, 1968).

58. G.W. Scott Blair, *Nature* **184** (1959) 354.

59. H.S. Souttar, *Brit. Med. J.* **69** (1925) 603.

60. A. Starr and M.L. Edwards, *Ann. Thorac. Surg.* **154** (1961) 726.

61. P.D. Stein and H.N. Sabbah, *Circ. Res.* **39** (1976) 58.

62. D.M. Stevenson and A.P. Yoganathan, *J. Biomech.* **18** (1985) 899.

63. D.M. Stevenson and A.P. Yoganathan, *J. Biomech.* **18** (1985) 909.

64. S.P. Sutera and M.H. Mehrjardi, *Biophys. J.* **15** (1975) 1.

65. K. Thalassoudis, "Numerical Studies of Flow Through Prosthetic Heart Valves", Doctoral Thesis, University of Adelaide, Australia, 1987.

66. K. Thalassoudis, J. Mazumdar, B.J. Noye and I.H. Craig, *Med. Biol. Eng. Comput.* **25** (1987) 173.

67. J.F. Thompson and Z.U.A. Warsi, *J. Comput. Phys.* **47** (1982) 1.

68. F.N. Underwood and T.J. Mueller, *A.S.M.E. J. Biomech. Eng.* **99** (1977) 91.

69. F.N. Underwood and T.J. Mueller, *A.S.M.E. J. Biomech. Eng.* **101** (1979) 198.

70. S.L. Weinberg, in *The Ureter*, ed. Bergman (Harper and Row, U.S.A., 1967) p. 48.

71. W.J. Yang and J.H. Wang, *Proc. 2nd Int. Conf. Mech. Med. Biol.* (Osaka, Japan, 5-7 June, 1980) p. 142.

72. W.J. Yang and J.H. Wang, *A.S.M.E. J. Biomech. Eng.* **105** (1983) 263.

73. A.P. Yoganathan, H.H. Reamer, W.H. Corcoran, E.C. Harrison, I.A. Shulman and W. Parnassus, *Artif. Organs* **5** (1981) 6.

74. D.F. Young, *J. Eng. Ind. Trans. A.S.M.E.* **90** (1965) 248.

75. K.L. Zierler, in *Handbook of Physiological Society, Vol. 1*, ed. W.F. Hamilton and F. Dow (American Physiological Society, Washington, U.S.A., 1962) p. 585.

INDEX